Conceptualizing Environmental Justice

Conceptualizing Environmental Justice

Plural Frames and Global Claims in Land Between the Rivers, Kentucky

Damayanti Banerjee

LEXINGTON BOOKS
Lanham • Boulder • New York • London

Published by Lexington Books
An imprint of The Rowman & Littlefield Publishing Group, Inc.
4501 Forbes Boulevard, Suite 200, Lanham, Maryland 20706
www.rowman.com

Unit A, Whitacre Mews, 26-34 Stannary Street, London SE11 4AB

British Library Cataloguing in Publication Information Available

Library of Congress Cataloging-in-Publication Data Available

ISBN 978-1-4985-0784-4 (cloth : alk. paper)
ISBN 978-1-4985-0785-1 (electronic)

∞™ The paper used in this publication meets the minimum requirements of American National Standard for Information Sciences—Permanence of Paper for Printed Library Materials, ANSI/NISO Z39.48-1992.

Printed in the United States of America

Contents

Preface

This book is a study of a conflict over environmental and cultural injustice in a rural community called the Land Between the Rivers in western Kentucky. Using participatory ethnographic field research, I study how the community mobilized their collective resources to pursue their protests against environmental and cultural injustice and seek legitimacy and support for their claims. This movement further reveals the theoretical implications of framing the notion of justice in environmental justice. I propose that environmental justice movements view justice as a large repertoire of protest themes and strategies that movement groups can choose from to articulate their claims.

Scholars of environmental justice have studied the disproportionate impact of environmental hazards and toxic wastes on low-income communities and minorities in the United States. However, perceptions of environmental injustice among affected groups are not limited to material inequities. Instead, communities facing unjust environmental actions view the pursuit of justice as a form of resistance to a broader spectrum of unfair practices, including cultural displacement and loss, lack of institutional access to seek justice, and capability building through political participation. In other words, perceptions of environmental injustice are not limited to a single aspect of justice. Movements often articulate a variety of issues and concerns and adopt these to mobilize protests, contest opposing claims, and negotiate solutions. The case of Land Between the Rivers is a fascinating study of competing claims and the processes through which these claims were fought, negotiated, and settled successfully.

In the course of my study, I examine how the dynamics of the central movement claims for justice were established over time, why specific claims gained prominence over others during conflicts, how residents mobilized to support and pursue those claims, and what solutions were negotiated after a period of struggle. I propose a four-dimensional environmental justice framework that includes conflicts over economic impact of displacement, cultural loss, institutional access and fairness, and political maneuvers for seeking alternatives. Each of these strategies were adopted and negotiated between contesting parties in the conflict and multiple and overlapping claims were also presented to seek solutions to conflicts. This does not imply that all the claims were successfully negotiated or gained equal support and resources. For example, the community was able to generate substantial support, political capital, and national attention to their resistance to an economic revitalization project, but had to rely on local resources for a church renovation project as part of their cultural justice claims. While both conflicts were successfully negotiated, the disparity in material support suggests unequal attention is often paid to what constitutes environmental injustice. This also raises critical questions in scholarly studies about how to articulate the contours of environmental justice and what constitutes a more inclusive framing of environmental justice. My ethnographic study explores these questions and proposes a critical roadmap of environmental claims that communities adopt to negotiate a fairer environmental system for themselves.

This book is a labor of love and a result of my decade-long ethnographic field study of an environmental justice movement in rural America, a part of the world that lacks sufficient representation in the environmental justice literature. I am indebted to numerous people who supported and guided me since I began this project. I am especially thankful to David Nickell, a former resident with deep family ties to Land between the Rivers and whose profound knowledge and love for his community and its environment has always impressed and inspired me to engage with and understand rural communities in the United States. He opened his home to me, escorted me around his community, introduced me to other residents, and answered my questions with great patience and understanding. This project would not have happened without his guidance and generous support. I am also grateful to Michael Bell, my dissertation advisor, whose great intellect is aptly matched by his humor, kindness, and a deep and abiding commitment to his students. He introduced me to Land Between the Rivers, taught me to think critically, and always encouraged me to appreciate the importance of both the social and the environmental in our lives. I am also grateful to David Pellow for his friendship and support and his patient readings and critical reflections on the many drafts of my papers on environmental justice. I am also grateful to Tammy Lewis and Dorceta Taylor for their encouragement and friendship. Finally, I am always grateful for the love, support, and patience of my family.

I wish to thank Ford Foundation Community Forestry Research Fellowship Program at the University of California, Berkeley for providing me with a generous fellowship to support my research. I also wish to thank the

Rural Sociological Society Endowment Grant Program, the Eugene Havens Award, and the Raymond J. Penn Scholarship at the University of Wisconsin-Madison for their financial support for my research project. I am grateful to Sarah Craig and Julia Torres of Lexington Books who were both patient and firm as they encouraged me to complete my book. I also appreciate my conversations with the U.S. Forest Service officials at Land Between the Lakes National Recreation Area, for providing me with a better understanding of the mission of natural resource conservation management programs in the United States. Special thanks to numerous community residents of Land Between the Rivers, who kindly invited me into their homes and indulged me with great kindness and patience as I sought to understand the nature and scope of the environmental movement in Land Between the Rivers. For that, I dedicate this book to the community of Land Between the Rivers.

As a scholar of environmental justice, I hope my book provides readers with a critical understanding of environmental justice *in action* as a set of strategic claims necessary for the pursuit of equity and fairness. More importantly, I hope it encourages us to collectively dedicate ourselves more firmly to seek and pursue environmental justice for all.

Part I

LOCATING ENVIRONMENTAL JUSTICE IN RURAL PLACES

Introduction

Locating Environmental Justice in Rural Places

Located in the far western corner of Kentucky, the community of Land Between the Rivers resembles tiny rural communities spread across the United States. Residents who called it home have lived in the neighborhood for generations dating back to the revolutionary war. Residents shared a long and eventful history in the region and formed enduring bonds of attachment to their place and community. In the mid twentieth century, the residents of Land Between the Rivers were displaced to pave the way for the construction of hydroelectric dams and a federally managed wildlife refuge. The displacement occurred as part of the New Deal, when Tennessee Valley Authority (TVA) was entrusted with economic development in the Southeast United States. TVA began the construction of the Kentucky Dam on the Tennessee River in the mid-1940s. In the early 1960s a second dam, Barkley Dam was constructed on the Cumberland River. As rising waters flooded the homes, residents were moved in five phases off the land until all were eventually forced out to make way for the wildlife refuge. It was renamed as Land Between the Lakes and promoted nationally as a success story for environmental education and economic development.

In response to this forcible displacement, the surviving residents began protests that shared important similarities with the environmental justice movement. While during the early years of the conflict the protesters focused solely on *cultural reparations*, their attention shifted to *economic injustices* as the community protested commercialization plans in the wildlife refuge in the mid-1990s. A third shift in the narrative emerged in mid-2000s as former residents of Land Between the Rivers pressed for greater *institutional participation in* policy decisions on cultural heritage and environmental protection. Finally, in recent months the former residents have organized parallel heritage tours of Land Between the Rivers as a *political protest* against the lack of transparency in decision-making. Thus, the study of the environmental justice movement in Land Between the Rivers suggests that environmental justice strategies are generally plural and evolving and are best conceived as a repertoire of strategies that movement groups adopt to mobilize their claims. While the protests have been long-standing with complex and evolving relationships between key actors, the following graphic shows the principal groups in Land Between the Rivers conflict. These groups include the former residents of the Land Between the Rivers community, regional hunting groups, anti-logging activists, national and local environmental groups, and the state management agencies, including the Tennessee Valley Authority, Fish and Wildlife Service, and the United States Forest Service.

The story of Land Between the Rivers is by no means a singular case study. As environmental justice movement spreads globally and gives voice to diverse discourses of justice, it has experienced an internal seismic shift.

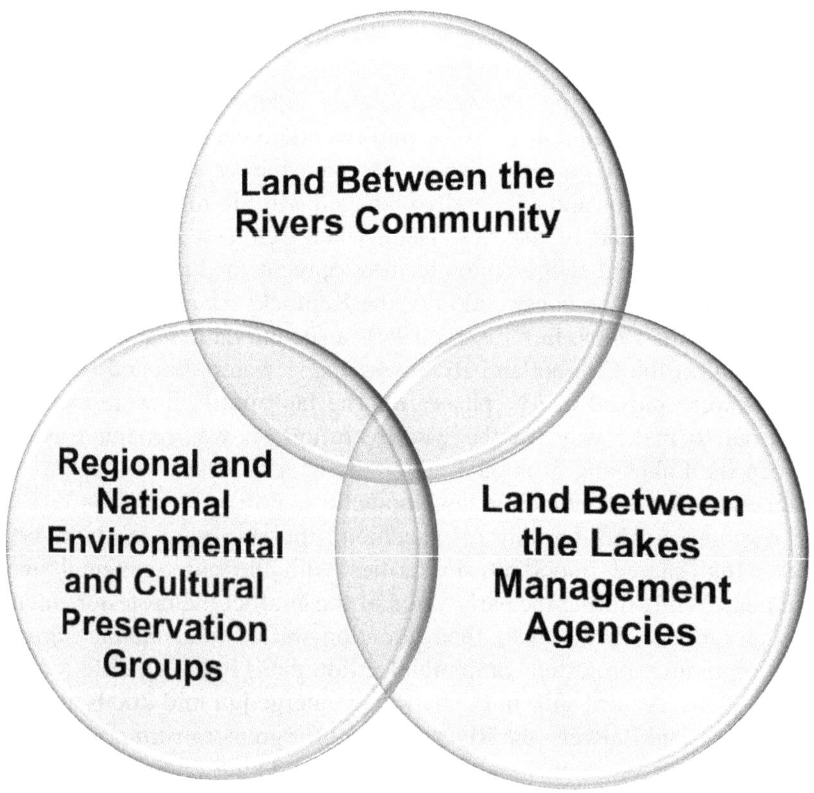

Figure 0.1 An illustration of the principal interest groups in Land Between the Rivers. Graphic courtesy of the author.

Environmental protests in the last decade encapsulate this transformation as activists discuss the importance of democracy, human rights, cultural identity, and political participation as integral elements of environmental justice. As movement groups organize communities against these injustices, they borrow from a wide repertoire of protest strategies to seek legitimacy and broaden their support. As environmental justice movements emerge globally, movement strategies are not limited to distributive/material concerns among communities, instead they include a diverse set of tools, including cultural loss, economic angst, as well as institutional transparency to mobilize communities.

Environmental justice movement in the United States draws our attention to the disproportionate impact of toxic deposits in communities of color and in poor communities. It is organized around the idea that all communities regardless of race, religion, or social class must have access to an equitable and healthy environment. The environmental justice movement supports communities affected by unjust environmental rules and provides a national platform to formulate and enforce environmental laws, regulations, and policies. Over the last two decades, environmental justice movements have enabled critical debates about racial injustices, income inequalities, human rights violations, cultural identity politics, and fair and democratic political participation. As movement groups mount protests against these injustices, they borrow from a wide repertoire of protest strategies to legitimize their claims and broaden their support.

As scholars of environmental justice, it is important to account for this diversity of justice themes in environmental justice movements. This book integrates and builds on these scholarly studies to propose a theory of environmental justice that argues for viewing EJ as a repertoire of strategies, or a "justice toolkit" used by communities to mobilize their resources and protest injustices. While researchers have examined many facets of justice over the years, scholarly research has consistently defined justice within specific disciplinary boundaries. In particular, three justice traditions have extensively addressed the contours of the notion of "justice" in environmental justice. First, environmental justice studies, a sub-field in sociology, explores the social justice struggles of the affected communities that are selectively targeted for disposal of hazardous industrial wastes. Scholars of environmental justice have explored prominent discourses within the field. Early scholarship explored how race as well as income determined discriminatory zoning practices in decisions about the siting of hazardous facilities (Anderton et al. 1994, Been 1994, Bullard 1994a, Mohai & Bryant 1992, Zimmerman 1993).

However, as scholars have noted, environmental injustices rarely emerge from discrete sets of overt actions (Pellow et al. 2000, Pulido 1996, Szasz 1997). In fact, environmental inequalities are also products of existing institutional mechanisms and ideological dynamics that produce unfair outcomes (Pellow et al. 2002). Environmental injustices thus can be explained through a critical analysis of socio-historical legacies of racism and classism that constrain life choices and deny important political and economic tools to actors for addressing such violations (Pulido 1996, Pellow 2000, Szasz & Meuser 1998). Similarly, as a movement deeply influenced by the key principles of the civil rights framework, the success of the environmental justice movement lies in its effective aligning of the civil rights paradigm with the environmental paradigm (Taylor 2000). This frame alignment, Taylor points out, helped in building coalitions between environmental, labor, and minority concerns and offered key political opportunities to successfully mobilize collective action. Overall, the environmental justice literature has examined inequities in the distribution and consumption of environmental "ills" and "benefits" as influenced by racial, economic, or political disparities.

Second, scholars in science and technology studies have framed the quest for justice as contestations between expert-guided "scientific" environmental knowledge on the one hand, and the local cultural knowledge claims on the other (Pickering 1992, Ottinger 2010). These knowledge contestations have occurred in the aftermath of environmental disasters, dam constructions, and other technological transformations. For example, science and technology scholars have studied the transformative impact of technologies on human-nature relationships with particular emphasis on the impact of constructing large technological systems like dams on poor and marginalized communities. Environmental protests against Narmada Dam construction in India, the Three Gorges Dam in China, and the Pak Mun Dam in Laos PDR provide rich resources for study of cultural contestations over scientific knowledge claims (Cheatham 2002, Sneddon & Fox 2006).

In the United States, TVA's first experimentation with dam building in Norris Basin, Tennessee, led to large-scale displacements and hardships as resettlement initiatives that experienced limited success and met with considerable resistance. Other studies have revealed knowledge contestations over dams in the Los Padres Dam Project in Brazil, where traditional beliefs, customs, and cultural knowledge aided in resisting scientific claims (Moore 1998).

Further, science and technology studies have also explored contestations over the nature of legitimate environmental knowledge, particularly, conflicts between expert and local/indigenous knowledge claims. For example, in the course of an anti-dam protest movement in Brazil, local communities have used local norms and cultural beliefs as tools to resist damming and claim cultural/spiritual relationships to their waters (McCormick 2006). Such knowledge contestations have inadvertently led to strategic cooptation whereby indigenous communities have refashioned and perfected scientific knowledge to stake their claims in environmental decision-making (Cohen & Arato 1994, Hess 2007, McCormick 2006). Further, knowledge contestations have all sparked interest in knowledge-sharing through citizen-scientist collaborations. Such collaborations are widely viewed as fundamental to ensure accountability in dam-building project policies and have occurred through information sharing, strategic co-ordinations, and research outreach programs (Allen 2003, McCormick 2006). Overall, justice in science and technology studies is advanced through contestations between scientific and indigenous knowledge claims.

Third, political philosophers have also contributed to studies of justice. In this view, justice is defined as emerging from a negotiated social contract whose success is deemed to be contingent on participatory equity for all (Sen 1981, Rawls 1971). Participatory equity refers to institutional fairness and citizens' ability to access and participate in institutional decisions that impact their well-being. For political philosophers, a just social order is predicated on the creation and availability of effective sociopolitical mechanisms and institutional structures that allow rival groups to participate and contest in political processes for facilitating justice. Thus, Amartya Sen (1999) states that successful implementation of a social justice contract must be contingent on the availability of social choices and democratic freedom to pursue those choices. Sen argues that these social choices, that include economic freedom, education, health care, and social safety nets, can only be accessed through the creation of equitable democratic institutions.

Thus for political philosophers like Sen, social justice can be pursued only through participatory politics advanced through a collective democratic contract among citizens. Sen's framework has been adopted widely by scholars of environmental politics. These scholars view the notion of participation as an important aspect of justice, one that guarantees a "place at the table" (Schlosberg 2007: 523). Drawing upon political philosophers these scholars see participatory justice as an important avenue for greater equity in environmental distribution.

Thus, within this stream of research, justice rests on the creation of and access to critical socioeconomic institutions that guarantee participatory fairness.

While the scholarly traditions cited above highlight specific facets of justice, in practice environmental groups rarely choose between one or the other forms of justice-seeking to organize their campaigns. Instead, movement groups draw upon diverse repertoire of strategies to mobilize their claims. For example, environmental and health concerns of workers in hazardous industries in the global south cannot be attributed simply to distributional inequalities. Instead, environmental injustices occur due to overlapping and complex factors that include weak environmental laws, unfair implementation of regulatory tools, unequal trade policies, lack of workers' representations, and poverty. Environmental movement groups acknowledge these challenges and adopt a plural set of strategic protest tools for mobilization against environmental injustice. Research on environmental justice must document these plural sources of injustice in order to account for the various forms of justice-seeking practiced within environmental movements.

This book distills and synthesizes the key themes from scholarly research to propose a theoretical framework for environmental justice. The opening chapters of the book synthesize the current literature on environmental justice and identify three justice themes in the literature. I build on these themes to propose a theory of environmental justice that conceptualizes justice as a plural and evolving set of strategies. The proposed framework includes four justice-seeking strategies: *cultural justice, economic justice, processual justice*, and *political justice*. The first three strategies are distilled from environmental justice research, science and technology studies, and political philosophy, respectively. Additionally, the fourth strategy, *political justice*, explores the means used by movement groups for constituency-building. Specific political justice tools include politicking, campaigning, lobbying, and appropriating established narratives to mobilize claims for justice, gain a public forum to seek support, and/or seek political coalitions for generating legitimacy (Banerjee 2015).

The book is divided into six main sections and includes seven chapters. The *Introduction* offers an overview of the book. Section I titled *Theorizing the "Justice" in Environmental Justice: A Plural Framework* reviews the field of research on environmental justice and proposes a four dimensional theory of justice that examines justice as a repertoire of strategies adopted by movement groups to mobilize their resources. The proposed theory builds on current studies in environmental justice, political philosophy, and science and technology studies and offers a four-dimensional framework that includes *cultural justice, economic justice*, *processual justice*, and *political justice*. The theoretical framework constructs a plural, multilayered, synthetic approach to justice that is indicative of the growing diversity in conceptions of justice within environmental justice movements.

In Section II *Theorizing the "Justice" in Environmental Justice*, I review the current literature on environmental justice to argue that definitions of "justice" in environmental justice must expand to include both distributive/material justice as well as non-distributive forms of justice. Specifically, I identify and examine the emerging themes of justice in the literature. I argue that environmental justice studies has examined the relationships between race, income, and environmental experiences in the United States and have argued that the impact of inequity in distribution of environmental resources unfairly impact poor and minority communities. Second, scholars in science and technology studies have explored a separate justice theme, one that examined how environmental contestations between expert-guided "scientific" environmental knowledge and the local cultural knowledge shape environmental conflicts.

These knowledge contestations are particularly evident in studies on environmental disasters and displacements over technological transformations among others. Finally, political philosophers have also contributed to studies of justice. These scholars have explored the institutional responses to demands for social justice. In this view, justice is defined as social contracts that are contingent on access to critical institutions that create the "promise of participatory equity" through greater institutional fairness (Schlosberg 2007, Sen 1981).

I then synthesize these themes of research to propose a theoretical framework for environmental justice. This framework conceptualizes environmental justice as a repertoire of justice-seeking strategies that include both distributive and non-distributive forms of justice. The flexible nature of this framework is indicative of the diverse and fluid nature of justice within environmental justice movements and offers to contribute to ongoing studies on the current possibilities and future directions for environmental justice studies.

In section III, titled *Researcher Statement on Field Study* I narrate the ethnographic experiences that shaped my research over several months in Land Between the Rivers. Ethnography offers an opportunity for story-telling. The role of the story-teller allows ethnographers to narrate their investment in their chosen research community as they negotiate their identities through the research process. As an observer immersed in her "field," the story-teller-researcher is engaged in analyzing her positionality in the course of her research. She is the stranger wearing multiple skins, a sojourner and a participant in the research process. In the course of this chapter, I draw upon Georg Simmel's notion of the stranger to explore how the stranger-ethnographer carves out her place in her story. Using my field notes and research experience in Land Between the Rivers, I examine the ethnographer's travels through her multiple "skins" or identities and explore how these embedded experiences shaped and influenced my study of the environmental justice movement.

In Section IV, titled *Just Places: A Story of Cultural and Environmental Injustice in Land Between the Rivers,* I examine the history of displacement in Land Between the Rivers. Located in between two rivers, Cumberland and Tennessee, Land Between the Rivers is an inland peninsula spread across the borders of western Kentucky and middle Tennessee. The residents of the region are the descendants of Revolutionary War veterans who settled in the area in the late eighteenth century. In the mid-twentieth century, Tennessee

Valley Authority confiscated the land using Eminent Domain and constructed two dams and a wildlife refuge in the region. The effect of the displacement on the community was two-pronged. First, management rights over environmental resources were transferred to the federal government. Second, the forced removal of the community led to a cultural displacement and loss of heritage for the residents. The dual nature of the loss is reflected in the ensuing movement which defined their struggle as a resistance to cultural and environmental injustices.

In the course of the next two chapters in this section I examine the story of a community whose settlement and eventual displacement led to a movement against environmental and cultural injustice. I begin with the history of the community settlement in the region in 1779 and trace the construction of a strong and abiding cultural identity and sense of place. Second, I examine the history of dam construction under the Tennessee Valley Authority as part of the socioeconomic development in the Southeast. Finally, I study the creation of a wildlife refuge and the transfer of community land under Eminent Domain in 1965. I explore how the displacement in five phases between 1945 and 1965 impacted the community's sense of history and cultural identity and shaped their protest movement against cultural and environmental violations.

Section V, titled *Contextualizing the Environmental Justice Movement in Land Between the Rivers,* narrates the emergence and evolution of the environmental justice movement in Land Between the Rivers with particular reference to the strategies adopted by the community over the course of five decades. I examine how the former residents of Land Between the Rivers adopted multiple injustice frameworks to gain support for their protests against the loss of their place and cultural heritage. Community protests were formally organized in 1965 when all land was confiscated under eminent domain. In the course of the five decades, the former residents have adopted various strategies to gain support and legitimacy for their struggle.

Early protests were largely directed toward experiences of *cultural injustice and stereotyping*. The movement reached an important landmark in the 1990s when *economic injustice* concerns emerged with TVA's decision to commercialize the refuge area. The post-1990s witnessed a transfer of administrative control from TVA to the U.S. Forest Service who abandoned TVA's commercialization plans returning the refuge to its original state of protected national forest. Buoyed by the success of their protests, the former residents pressed forward and the movement entered a new era with greater emphasis on *processual justice* and community participation in environmental and cultural decision-making in Land Between the Lakes. The recent setbacks in gaining institutional justice however have led the movement in a fourth direction. The movement is once more entering the public domain through its adoption of *political justice* strategies including demand for signatory status to boost effective participation in environmental decision-making, initiating new cultural heritage projects to counter the "official" plans for heritage management, and conducting community-guided parallel heritage tour of the grounds.

Section V titled, *Towards an Integrative Framework for Environmental Justice* is a concluding account of the book and discusses the key implications of the four-dimensional environmental justice theory for future research on environmental justice. In this section, I examine the lessons learned from the environmental justice movement in Land Between the Rivers. In particular, I explore two possible implications of the proposed theoretical framework for future research on global environmental justice movements. First, the proposed theory allows us to conceptualize environmental justice as a synthetic theory that views justice as more than a question of distribution. In fact, environmental justice movements draw upon a diverse repertoire of strategies to gain support and legitimacy for their movement. As communities dealing with disaster rebuild their lives and homes, their diverse needs cannot be captured only with a distributive lens. Instead, studies show that affected communities incorporate multiple protest strategies to demand better support as they rehabilitate and rebuild their communities. A multivalent, synthetic framework allows us to incorporate these diverse experiences in our framework.

Second, studies of global environmental justice movements are increasingly focusing on studies of life-cycles of toxic products from its productions to its disposal (Pellow 2007). As movements emerge, coalesce, formalize, and succeed or fail, movement groups adopt and legitimize the use of diverse strategies of protest (Tilly 1978). Land Between the Rivers is a case in point. While early protest insisted on cultural reparations, the movement was formalized and coalitions emerged as the community activists adopted and prioritized their claims against *economic injustices* during the commercialization era. Important strides during this phase also facilitated in lobbying for institutional justice and participation. Finally, limited transparency in decision-making and the community's decision to keep up the pressure have led to adoption of parallel political initiatives for justice. A life-cycle approach to such stories of environmental justice allows us to integrate these diversities into the environmental justice framework.

Part II

THEORIZING THE "JUSTICE" IN ENVIRONMENTAL JUSTICE

Chapter 1

An Integrative Framework for Environmental Justice

Environmental justice movement began in the United States in 1970s in response to widespread concerns about impact of toxic pollution in minority and poor communities. The movement built on the principles of the Civil Rights and Labor Movement and sought grassroots actions against environmental inequalities. Scholars within this tradition have explored the factors that caused the production and distribution of environmental waste as well as the impact of the disproportionate burden of toxic contamination among vulnerable communities. The Environmental Justice movement has been extensively documented in academic literature and has helped in the emergence of Environmental Justice Studies as a field of research (Agyeman 2003,[1] Bryant & Mohai 1992,[2] Pellow 2004,[3] Taylor 2000,[4] Holifield et al. 2009[5]).

Despite these developments, environmental justice studies have focused predominantly on material/distributive injustices as the primary drivers of environmental injustices. Some of the seminal studies on environmental injustice have documented displacement due to dam constructions in Brazil (McCormick 2006), public health impact of toxic dumping (Bullard 1993, Bryant & Mohai 1992), and workers' rights issues among ship-breaking workers in India (Demaria 2010). These studies draw our attention to how certain groups and communities have come to bear a disproportionate brunt of environmental hazards and/or have been excluded from sharing the benefits. In addition to these material injustices, scholars have also explored non-material concerns—including cultural, political, and human rights issues—in the course of environmental conflicts. For example, the destruction of a particular ecosystem, a way of life, or cultural traditions at the hands of the state or corporations involved in pollution and/or resource extraction might put an end to various cultural practices and significant social symbols, in addition to harming various material objects. As such, these outcomes are not always easily quantifiable or distributable and therefore they fall outside the realm of the material basis of many environmental justice movement claims and Environmental Justice scholarship, both of which tend to rely on a distributive model of justice.

Non-material forms of injustices are well-documented within the larger environmental studies and offer exciting opportunities for cross-fertilization of ideas. For example researchers within science and technology studies and political philosophy have been actively engaged in documenting environmental justice movements. Science and technology studies provide an effective theoretical framework for incorporating cultural and emplaced narratives of environmental justice stories. Further, political philosophical writings on issues of fairness and equitable access to institutional justice through sustainable policymaking offer important tools to seek justice for disaffected communities. Hence, the heterogeneity of discourses on the notion of "justice" in environmental justice, as evident in these scholarly traditions suggests that the robustness and conceptual métier of the justice frames lies in its plurality of vision and perspectives.

This chapter synthesizes these complementarities of approaches and proposes a four-dimensional framework of environmental justice. I begin with a review of the scholarly approaches to environmental justice and identify the underlying commonalities in these definitions. Second, drawing upon these overlapping theoretical frames, I propose a four-dimensional approach to environmental justice—*cultural justice*, *economic justice*, *processual justice*, and *political justice*. I argue that environmental justice movements are best viewed as a repertoire of justice tools, and mobilization efforts often incorporate one or more of these protest strategies. I further argue that the choice of protest strategies is largely context-driven and temporally bound and may alter over the years. Further, in the next few chapters I apply this framework to examine how community mobilization occurred in the course of an environmental justice movement in a community called Land Between the Rivers, Kentucky.

DISCOURSES ON ENVIRONMENTAL JUSTICE: A REVIEW OF THE LITERATURE

In *Silent Spring* (1962) Rachel Carson presented one of the early environmental justice studies exploring the impact of chemicals and other toxins on our environments. Carson examined the destructive ecological forces at work in modern society, particularly in our overuse of chemicals and pesticides like DDT. Her work explored the poisonous effects of chemicals on nature that ultimately compromised our environments and the greater community well-being. As public concern grew over the long-term implications of industrial pollution, activists and scholars alike devoted greater attention to the ramifications of toxic pollutants on neighborhoods.

In the late 1970s, the Love Canal movement in Niagara Falls raised key questions about environmental health, community well-being, and corporate responsibility. The grassroots protest movement in the working-class neighborhood of Love Canal came to symbolize the looming environmental disaster represented by industrial pollutants in the United States and paved the way for the rise of the environmental justice movement (Blum 2008). Similar protests against siting of toxic wastes followed in communities across the United States. In Warren County, NC, a predominantly African American community resisted the siting of PCB-contaminated soil in their neighborhood and organized sit-ins and other nonviolent protest strategies (Labalme 1988). The peaceful protest strategies, reminiscent of the civil rights movement, were instrumental in highlighting the systemic relationship between poverty, race, and vulnerabilities to environmental risks.

Environmental justice studies define justice as the process by which communities aim to ensure *fair and equitable distribution of environmental goods and bads through grassroots action and policymaking*. Scholars within this tradition examine the social structural inequalities that contribute to an uneven distribution of resources and the impact of resource inequity on poor and minority communities. The framing of environmental issues in terms of "racism" and "injustice" suggested a radical shift from traditional environmental activism that conceptualized nature predominantly as pristine wilderness. For affected communities, however, the environmental injustice frame resonated clearly as a tool for understanding and resisting environmental challenges. It also offered an important conceptual lens for scholarly research on environmental justice. Since the early 1980s, scholars of environmental justice have documented the connections between distributional inequalities in environmental resources and its unequal impact on marginalized communities (Bullard 1983, Scott 2010). Broadly, these scholars examine the influence of race, class, gender, and global geopolitical inequalities on distributive decisions about industrial and other environmental hazards.

Scholarly interests in questions of environmental inequality date further back. Empirical evidence of the disproportionate impact of economics on environmental well-being was already considered in the 1970s by both geographers and economists (Berry et al. 1977, Burch 1976, Freeman 1972, Zupan 1973). While these studies offered important insights about poverty and environmental risks, little attention was paid to examine the causal impact of race on toxic exposure. Studies on distributional inequalities leading to environmental injustices were explored by the Governmental Accounting Office in 1983 (GAO 1983). Other similar findings were reported that documented similar forms of inequalities (UCC 1987; Bullard 1983; Bryant & Mohai 1992; Bullard 1994; Freeman 1972; United Church of Christ Report 1987).

Other studies conducted to measure the impact of social inequities on environmental experiences of marginalized communities explored perceptions of environmental racism among Native Americans and Mexican/Chicano Americans (Berry 1977; Bath, Tanski, and Villareal 1994; Comancho 1998; Hooks and Smith 2004), impact of toxic releases (Bowen et al. 1995), occupational exposure (Friedman-Jimenez 1989), waste facility siting (Been 1993), and unequal enforcement (Lavelle and Coyle 1992).

Emergent themes within the environmental justice scholarship specifically examined how inequities in environmental distribution affects marginalized groups, the institutionalized nature of unjust distribution, and its impact on framing environmental justice conflicts and charting the course of environmental activism (Pellow 2004; Taylor 2000). Additionally, researchers have examined place specificities in conceptions of environment, cultural politics in conflicts over environmental distribution, and political economic inequities in environmental siting decisions (Adamson et al. 2002, Agyeman et al. 2003, Marbury 1995, Martinez-Alier 2003). Researchers have also examined how current discourses on climate change need to be informed by distributive and institutional barriers to a cooler planet (Checker, 2010, Dorsey, 2007, Roberts & Park 2009, Stallworthy 2009). Overall, these studies found that racialized policies and enduring material inequities are key determinants of unequal distribution of commercial hazardous waste facilities and other environmental pollutants.

While the prominent studies on environmental justice during the early years offered empirical insights about environmental inequalities in poor and minority communities, more work was needed to understand the socio-historical mechanisms or processes of distribution that contributed to marked inequities (Capek 1993, Pellow 2000, 2004, Pellow et al. 2002, Pulido 1996, Szasz and Meuser 1997, Taylor 2000).

Hence, in the mid-1990s, scholarly focus on the institutional mechanisms and processes of distribution led to a creative surge in the scholarly contributions (Szasz & Meuser 2000; Pellow et al. 2000; Taylor 2000). Scholars explored the historical genesis of the racialized environmental policies toward marginalized groups in the United States. Thus, in their study of environmental inequality in Santa Clara County in San Jose, California, Szasz and Meuser (2000) pointed out that "any attempt to depict industrial siting as the imposition of undesirable development on certain victimizable communities, in spite of misgivings and opposition in those communities, is an anachronistic projection of contemporary attitude" (Szasz & Meuser 2000: 15). Instead, they argued that such association needs to be viewed as products of broader processes of racialization that determine and shape occupation, financial assets, and general life opportunities. They emphasize the embedded nature of racial discrimination in the historical experiences of minority communities and suggest that mere intentionality cannot explain this complex process.

Other key contributions include seminal studies on enduring political processes that shaped American environmental policies. Pellow's (2000) groundbreaking theoretical framework "Environmental Inequality Formation (EIF)" explored the specific socio-historical, institutional, and political processes, by which environmental inequalities are created, reproduced, and challenged (Pellow 2000, Pellow et al. 2002, Pellow 2004). Pellow's framework explored the impact of multiple social processes and actors in shaping uneven and unjust environmental policies. For Pellow, environmental inequities result from actions and interactions of multiple actors including the state, corporations, environmental groups, residents, and churches, whose cross-cutting interests and overlapping allegiances shape the environmental discourse.

Similar studies on theorizing the nature and scope of environmental justice analyzed the construction of an environmental justice frame to garner grassroots support in community mobilization. Taylor (2000) ably demonstrates effective adoption of frame alignment strategies in environmental justice movements. She argues that the success of the environmental justice movement rests its ability to link environmental injustice claims to the master frame of civil rights movement. Other scholars, like Pulido (1996) argue that predominant environmental justice literature focuses too much on the statistical importance of race thereby providing a simplistic analysis. She postulates that by underscoring overt racism as contributing to dumping, environmental justice literature fails to address racism as an ideology. Pulido questions the very rationality of a debate where racism is reified as an independent attribute that is identifiable through a study of a series of discrete actions of racism, while ignoring the ideological and institutional dynamics that shape such actions (Pulido 1996).

In recent years, researchers have examined how the global spread of the environmental justice master frame has shaped the movement's scope and message. Scholars have explored the political economy of garbage production and transfer to the Global South creating new pollution zones in the process with important implications for a globally risk society (Beck 1992; Marbury 1995). As environmental justice claims emerge globally, it will seek to accommodate plural claims of justice. As Walker (2009) aptly observes the elasticity of the environmental justice frame shows its capacity to evolve and recontextualize itself as it expands horizontally and across multiple vertical scales. Urkidi and Walter (2011) refer to these multi-scalar approaches to environmental justice in anti-gold mining justice movements and argue that movement actors jump scales as they attempt to align their movement frames with national and global political processes.

As environmental justice movements adopted the plural iterations of justice concerns, the conceptual framework is increasingly expanding beyond traditional analyses of global redistribution patterns and engaging in a "diversity of unequal interactions between multiple environments and multiple forms of social difference" (Walker 2009: 375). As environmental justice movements expand globally, the scope and nature of injustice concerns have proliferated. In particular, key environmental justice concerns now include patterns of cultural contestations over environmental justice claims, human rights, and institutional equity in environmental distribution, global waste generation and transfer, new space of pollution, engendered environmental justice, and climate justice (Ukridi and Walter 2011, Schlosberg 2007, Demaria 2010, Pellow & Brehm 2013, McCormick 2006).

A second scholarly tradition studies how solutions for environmental justice depend on creating more equitable access to democratic institutions. These scholars examine how socioeconomic and environmental policies contribution to capability building among the poor empowering them to choose the necessary entitlements for their individual and community well-being. These scholars have defined justice as contingent on institutional equity, democratic transparency, and participatory fairness. Justice, in their articulations, emerges from debates over the ability to access institutional resources and participate in institutional decision-making. These justice approaches build on key studies in political philosophical literature on social justice and extend them to analyses of environmental justice conflicts. Thus urban geographers have explored the urban parks movement as an example of more equitable green spaces, while cultural theorists have explored the role of the racialized state capabilities building (Walker 2009; Buckingham et al. 2009; Kurtz 2009). Still others have explored the discursive politics of environmentalism to redefine environmental justice (Holifield 2009).

Finally, environmental justice theorists have drawn upon political philosophical traditions to argue that distribution, recognition, and participation play an import role in framing and mounting environmental justice movements. Thus, activists resisting environmental health violations in the shipbreaking industry in the Global South have highlighted the relationship between lax implementations of environmental policies, lack of legal recourse for affected workers in the industry, and poor institutional support systems as important challenges. This notion of procedural and institutional justice has been adopted in scholarly studies of environmental justice. Scholars like Schlosberg postulate the need for critical engagement and dialogue among various interest groups as key to institutional justice (Schlosberg 2004: 523).

Finally, a third theoretical tradition examines how perceptions of injustice shape *feelings of cultural loss and community identity*. These studies emphasized the increasing *enculturation and democratization of institutional scientific knowledge* as it is co-opted by displaced and devastated communities to rebuild their places and cultures. These compelling narratives tell us stories of heterogeneous citizen-science alliances as communities increasingly adopt scientific knowledge to counter state and corporate environmental rhetoric. David Hess examines the generative role of environmental movements in adoption of new forms of technologies and the possibilities for success for these movements (Hess 2007).

Scholars in Science and Technology studies of the environment have also studied contestations between local cultural knowledge and expert knowledge systems in order to examine how knowledge gaps in expert structures contribute to health epidemics in the aftermath of environmental violations (Frickel & Vincent 2007; Frickel et al. 2009). Allen examines the conflicts over the legitimation of contesting knowledge claims in the aftermath of Hurricane Katrina as the city began its reclamation and rebuilding initiatives (Allen 2007, 2008). These debates over democratization of science also find expression in Ottinger's study of Norco, Louisiana (Ottinger 2010). While acknowledging the potential of a citizen-science approach for democratizing environmental policies, she critically reflects on its effectiveness in influencing scientists and decision makers as scientific standards set for data monitoring act as both bridge builders and policing mechanisms. Her paper calls our attention to the need to revisit standard setting as an important site of intervention in democratizing scientific knowledge (Ottinger 2010). This emphasis on democratizing expert knowledge by building citizen-scientist alliances provides an effective lens in rebuilding narratives in disaster areas.

These heterogeneous perspectives on justice highlight specific facets of justice explored within three scholarly traditions. However, in practice, environmental groups rarely choose between one or the other forms of justice-seeking to organize their campaigns. Instead, movement groups draw upon diverse repertoire of strategies to mobilize their claims. For example, environmental and health concerns of workers in hazardous industries in the Global South cannot be attributed simply to distributional inequalities. Instead, environmental injustices occur due to overlapping and complex factors that include weak environmental laws, unfair implementation of regulatory tools, unequal trade policies, lack of workers' representations, and poverty. Environmental movement groups acknowledge these challenges and adopt a plural set of strategic protest tools for mobilization against environmental injustice. Research on environmental justice must document these plural sources of injustice in order to account for the various forms of justice-seeking practiced within environmental movements.

This chapter integrates these scholarly approaches to propose a theoretical framework of environmental justice. Below, I begin with a review and synthesis of the current literature on environmental justice and identify three justice themes in the literature. Second, I propose a four-dimensional theoretical framework that explores how environmental justice movements use plural justice-seeking strategies to mobilize their claims.

Justice as Distributive Fairness: Environmental Justice Approaches

The pursuit of social equity for the most vulnerable sections of society has remained an important concern in studies of environmental justice. While the early literature on environmental justice primarily focused on questions of unequal distribution of environmental "bads," recent years have witnessed a burgeoning concern for examining the globalizing trends in environmental justice activism, reflecting the trajectory of environmental justice movements.

Three studies, considered the precursors to the American environmental justice movement, have emphasized the importance of unequal distribution in creating and sustaining environmental injustices. To begin with, the General Accounting Office report, studying hazardous waste site locations in the United States, found that three out of four landfills are located near predominantly African American communities (GAO report 1983).

A second landmark study conducted by the United Church of Christ (UCC) in 1987 using Zip codes found that 37.6 percent of landfills are located near predominantly African American neighborhoods in the United States (UCC report 1987). The UCC study was particularly important because it was the first comprehensive attempt to present a national analysis of data on landfill locations that made an explicit connection between race and the likelihood of exposure to toxic waste and other environmental hazards (Taylor 2000: 28). A third classic study was that of Robert Bullard, who found that 21 of Houston's 25 waste facilities are located in African American neighborhoods (Bullard 1983).

All three studies found that race is the single best predictor of unequal distribution of commercial hazardous waste facilities and other environmental bads (Mohai and Bryant 1992). Similar arguments were presented in a range of case studies that followed on the impact of environment racism on Native Americans and Mexican/Chicano Americans (Berry 1977; Bath, Tanski & Villareal 1994; Comancho 1998; Hooks & Smith 2004). There have also emerged studies on toxic releases (Bowen et al. 1995), occupational exposure (Friedman-Jimenez 1989), waste facility siting (Been 1993), and unequal enforcement (Lavelle and Coyle 1992) which further challenge faulty distribution practices in creating unequal environments.

In subsequent years, incidences of large-scale environmental disasters broadened the conceptual axis of justice to address *social inequalities,* focusing on both racial and class-based distributive inequalities. Scholars examined both patterns of distribution and mechanisms or processes of distribution that contributed to marked inequities (Capek 1993, Pellow 2000, 2004, Pellow et al. 2002, Pulido 1996, Szasz & Meuser 1997, Taylor 2000). For example, in their study of environmental inequality in Santa Clara County in San Jose, California, Szasz and Meuser (2000) pointed out that "any attempt to depict industrial siting as the imposition of undesirable development on certain victimizable communities, in spite of misgivings and opposition in those communities, is an anachronistic projection of contemporary attitude" (Szasz & Meuser 2000: 15). Instead, they argued that such associations need to be viewed as products of broader processes of racialization that determine and shape occupation, financial assets, and general life opportunities.

Others like, Pellow (2000) adopted a more institutional approach to environmental justice theory and examined the historical and institutional processes that create, reproduce, and sustain environmental inequalities (Pellow 2000, Pellow et al. 2002, Pellow 2004). In a case study of the construction of a toxic incinerator in the African American neighborhood in Robbins, Illinois, Pellow (2002) examined the factors that explain why a community itself was actively recruiting an incinerator. He argues that an adequate explanation rests on a shift from perpetrator-victim analyses to a critical examination of the political economic processes of production and distribution under globalization. This alternative understanding, Pellow argues, will help us expand the scope of analyses of environmental racism from community-specific, "localized" studies in the United States to a much more broadly based "global" analysis of transnational trade of hazardous waste under the guise of "garbage imperialism" (Marbury 1995, Pellow et al. 2002).

A third approach uses frame analysis to examine how environmental justice movements were able to build on a groundswell of support to achieve considerable success in the United States (Capek 1993, Epstein 1997, Taylor 2000). Taylor (2000) develops the notion of what she calls the *environmental justice paradigm* (EJP), which primarily attempts to theorize the social construction of environmental problems. Drawing upon the framing literature, Taylor (2000) argues that the astounding success of environmental justice movement was primarily due to frame alignment with the civil rights movement discourse. Thus, Environmental Justice movement's ability to form strategic coalitions with the civil rights movement helped it gain legitimacy among communities affected by environmental health challenges.

Justice as Rights: Entitlements and Capabilities Approach

More recent theoretical approaches to environmental justice draw closely upon political philosophical traditions. Political philosophers have traditionally conceptualized justice primarily in terms of equality and human rights. Scholars including Steven Lukes, John Rawls, and Amartya Sen are instrumental in providing sophisticated models of justice while differing in their conceptual details. Thus for example, in his *Five Fables of Human Rights* (1993) Steven Lukes defines justice as a rationalizing and essentially political discourse. He sheds light on the centrality of human rights in definitions of justice by asking us to imagine five types of society following the Weberian ideal types. "Utilitaria," his first ideal type, rejects human rights for it stands counter to the sovereignty of the collective. The second ideal type, what he calls "communitaria" finds human rights discourse unsuitable since it presents a set of abstract principles that is at odds with their concrete way of life. Similarly, "proleteria" the third type of social system rejects human rights since it softens hearts against class enemies and is detrimental to class interests.

For Lukes, only two forms of ideal typical societies can adequately conceptualize justice and human rights, though they differ in their implementation. "Libertaria," his next ideal type, conceives human rights in terms of market freedom and property rights. However, libertarian economic system is based on a narrowly defined self-interest where, as Lukes points out "the individual is impervious to the thought that the others have more urgent claims on resources, or that some of his advantages are gained at the expense of others" (Lukes 1993: 33). Thus for Lukes, an effective implementation of human rights requires that basic forms of equalities be taken into account which is only feasible when a society transcends self-interested goals. Hence, Lukes argues that a successful application of the principles of justice and human rights is only possible in his final ideal type, what he calls, "egalitaria" which suggests a positive commitment to an equal access to civil rights as well as a general striving toward growth and social improvement.

Lukes' framework is useful in exploring the necessary political arrangement that best promotes the realization of the principles of social justice. Yet, his approach remains primarily foundational in nature with little attention to questions of power. In other words, his framework is a quest for a notion of universal human culture with little or no analysis of power relations in everyday social and political discourses. Thus, while Lukes' model of human rights underscores the importance of principles of equalitarianism, he does not address the conditions necessary for achieving it. Hence, his framework does not account for power differentials within societies that influence negotiations over human rights. The foundational basis of Lukes' theory has been critiqued by John Rawls, who suggests a more politically embedded notion of justice (Rawls 1971, 2001).

Rawls's theory of human rights postulates a minimum political standard necessary in every society for a successful implementation of justice principles. He introduces the notion of the "original position" whereby people are fairly represented and rational and are engaged in a reflexive debate over principles of human rights (Rawls 1971). Unlike Lukes, Rawls's framework is more open to political positions, dialogic relations, and reflexive engagements. Concepts of justice and human rights emerge in Rawls's work from this position of reflexivity. What is unique to Rawls is his emphasis on "recognizing" the possibilities for building a society based on equalitarian principles (Rawls 2001; Schlosberg 2007). Yet, the Rawlsian model does not grapple with questions of power, social inequality, and distributive justice—issues that are at the forefront of environmental justice theory. And since we do not conceive ourselves as occupying the original position and wearing a "veil of ignorance" (Rawls 1971) in our everyday lives, it is difficult to conceptualize justice theory *sans* a discourse over power relations.

In conceptualizing his theory of social choice and individual functionings, Amartya Sen, an economist and a political philosopher, addresses the limitations of the Rawlsian model. In the course of exploring questions of development and social justice, Sen points out that fair distribution of resources remains ineffective unless accompanied by people's freedom to choose the resources they want, depending on the contexts in which they live. The pursuit of principles of fairness and freedom can only occur by building people's capabilities, which in turn requires a social order based on principles of dialogue and democracy. Sen's approach to a theory of justice is twofold. First, a theory of justice needs to begin with the creation of equitable principles, and second, it creates provisions for effective social choices that allow us opportunities for realizing our *functionings*. Social justice, in Sen's view, links questions of distribution to creation of social choices. Social choices are only valued when opportunities for *capability formation* are available for accessing the choices. Capability formation thus is the essence of Sen's concept of social justice and is constitutive of material/distributive fairness, availability of a broad range of choices (both material and non-material), and an equitable political platform necessary to attain both (Sen 1999, 2000).

Sen defines capability formation as primarily an indicator of an individual's well-being based on her *functionings*. Sen defines functionings as a person's capacity to access available resources to improve his/her life (Sen 1999). Hence, an individual's ability to reach his/her functionings is dependent on the availability of a range of choices that will improve his/her life. Achieving this capability suggests that one is "free to achieve alternative functioning combinations or freedom to achieve various alternative lifestyles" and that injustice results from the non-availability of choices (Sen 1992: 46). Capability formation thus moves the discussion away from the type of goods available for our use to a debate on what these goods can do for our well-being (Sen 1992)

Each of these approaches highlights the need for a multidimensional understanding of justice. While Lukes calls for a need to reexamine the foundations of society within which justice is negotiated, Rawls argues for a clearer analysis of what constitutes a fair contract among participating actors. Sen on the other hand, draws our attention to the impact of sociopolitical constraints and inequities of power on the creation of a just social system. Just as environmental sociologists examine the causes, processes, and consequences of environmental inequalities, political philosophers theorize about the constitutive nature of social justice—complementary concerns that shape and reinforce each other.

David Schlosberg's efforts to redefine the "justice" of environmental justice draw closely upon political philosophical traditions (Schlosberg 2004; 2007). He suggests that theory development in environmental justice should closely draw upon global environmental movements. Schlosberg argues that environmental justice movements frequently transcend distributive questions and offer a more plural reflexive framework of justice. Hence, a theory of environmental justice must account for a threefold approach to justice: equity in the distribution of environmental risk, recognition of the diversity of the participants and experiences in affected communities, and capability building for participation in the political processes (Schlosberg 2004). Schlosberg's work attempts to transcend the logic of distributive justice to focus on analyses of processes, procedures, and recognition of inequalities in environmental outcomes (Schlosberg 2007). Schlosberg finds close parallels between the capabilities approach (proposed by Amartya Sen) and environmental justice discourses. The latter, he argues incorporates visions of community capacity building to emphasize the need to value community health and well- being.

Further, in a recent ethnographic study of anti-gold mining justice movements, Urkidi and Walter (2011) adopt Schlosberg's three-dimensional justice framework to chart the course of an organized environmental movement. Their paper finds evidence of the use of a multivalent framework of justice much like political philosophical propositions. They argue that movement actors jump scales (from participation, to recognition to distribution) as they attempt to align their movement frames with national and global political processes (Urkidi & Walter 2011).

Finally, in a recent paper Harrison (2011) draws upon Steven Lukes' five-dimensional justice framework to examine an ethnographic case study of political strife over agricultural pesticide drift in California. She argues that the environmental inequalities of pesticide drift stem not from a lack of justice, but from adoption of various definitions of justice including libertarianism, utilitarianism, and communitarianism among others.

Justice as Encultured Democracy: Science and Technology Studies Approaches

Scholars working at the intersections of science and technology studies (STS) and environmental justice make a two-pronged contribution to theorizing the "justice" in environmental justice—*democratizing scientific knowledge* and *incorporating local cultural knowledge* into scientific discourses. These are not mutually exclusive categories and find expression in scholarly discourses within STS studies. Thus for example, STS scholars argue that environmental justice narratives foreground the debates over the primacy of local versus expert knowledge claims as disaster-affected communities attempt to rebuild their places. In her ethnographic study of Hurricane Katrina, Allen (2007) points out the politics of rebuilding in New Orleans as questions about whose science and knowledge claims "counts" emerges as an overarching concern. As New Orleans slowly returns to its former self amid reclamation attempts, it has added new discourse strategies of justice to the repertoire of environmental justice. Allen (2007) suggests that citizens' groups are building multilevel alliances with organized national and international agencies as it seeks out novel relationships with independent scientific experts and professional groups. Allen suggests that such heterogeneous alliances are strong indicators of a growing democratization of science as citizen-scientists learn and adopt scientific strategies to counter state and corporate narratives (Allen 2007).

Others like Simms (2007) point toward the unique nature of Hurricane Katrina as a social and technological disaster (in addition to a natural one). Such disasters allow us to reflect on the network of social and material relations that inform and shape use of technologies in our everyday lives (Simms 2007: 94). Technologies inform our social infrastructures and normalize risks by enforcing its invisibility. We increasingly live in a world defined and shaped by technological transformations of nature that leave us vulnerable to environmental risks, infrastructure failures and more broadly to Katrina-like "normal accidents" (Perrow 1982, Freudenberg 2009, Simms 2007). Such technological disasters also open up possibilities for reformulation of scientific knowledge by building effective citizen-science coalitions through democratizing scientific knowledge. Such attempts however remain ineffective with inadequate institutional support and as Stevens (2008) points out can be better supported by redefining the role of the scientist as a public intellectual, incorporating local knowledge claims into scientific discourses and consistent shaping of research and data with nonacademic audiences. Fischer (2000) examines the potential consequences for democratization of deliberative structures by encouraging meaningful non-expert (citizen) engagement in environmental policymaking. He argues that such alliances will provide relevant local knowledge necessary for successful policies.

Still others call for a reexamination of the conceptual contours of citizen science for the effective realization of its transformative potential. Thus for example, Frickel and Moore (2006) argue that an in-depth study of the social and institutional disparities and the consequences of competing knowledge claims can be better explored within a "political sociology of science" framework. Such a framing will allow scholars to examine the successful co-optation of scientific discourses for collective action frames (Epstein 1996, Lerner 2005), the creation and sustenance of a public sphere for community deliberation on science (Beck 1992, Irwin 2001). Critiquing the need to reassess standard setting as an important site of intervention, Ottinger (2010) argues that the contradictory role of current scientific standards as conducting both boundary bridging and boundary policing responsibilities drastically reduces the effectively of citizen science (Ottinger 2010: 3). Instead, she points out, a truly participatory democratic citizen science rests on the successful transformations of institutional policies regarding scientific standards.

The Plurality of Justice Narratives: An Environmental Justice Framework

Justice, in the parlance of environmental justice is defined as an analytic concept to examine and understand how resources are distributed unequally with deeply divergent impact on communities. As discussed earlier, scholarly traditions have long debated the nature and scope of the "justice." Political philosophers view justice as incumbent on the creation of a set of institutions that directly or indirectly allocate liberties, rights, and other material and non-material resources to its members (Miller 1999: 152). Political philosophers refer to this as the question of *entitlement* whereby an existing political community has the conditions necessary for equal access to the resources available to them. STS scholars, on the other hand, argue that effective distribution rests on democratizing scientific knowledge claims by refashioning deliberative structures in society and by incorporating local knowledge systems into scientific discourses (*local place and democratic principle*). Finally, environmental justice studies have adopted and applied the notion of justice to view how environmental resources are distributed and the interactions between race, class, and environmental experiences of marginalized groups in the United States and around the world.

Below I closely draw upon these intellectual commonalities and critical divergences to propose a four-dimensional approach to environmental justice, that include—*cultural justice*, *economic justice*, *processual justice*, and *political justice* (Banerjee 2014). First, *economic justice*, our first dimension, is well-documented in environmental justice and refers to conflicts over access to material resources among contesting communities and groups. Second, science and technology studies scholars have argued that *cultural justice*, our second dimension, offers an important tool for narrating how places shape environmental politics and experiences of injustice. Environmental justice conflicts occur in place. Places are constructed through shifts in political and cultural locations during environmental conflicts and through processes of negotiations whereby cultural meanings, political identities, social values, and symbolic perceptions are internalized within an environmental discourse which in turn informs social relationships.

Third, political philosophers have successfully conceptualized the notion of *processual justice,* our third dimension. Processual justice refers to institutional arrangement within which environmental negotiations are initiated among contesting actors. Finally, *political justice*, our fourth dimension, refers to the act of politicking that movement groups and communities engage in to legitimatize their cause and mobilize resources for success. Each of these dimensions find expression in the story of Land Between the Rivers as the community engaged in resisting the loss of their cultural and environmental rights due to dam-building. While, the next chapter applies this theoretical framework to the environmental justice movement in Land Between the Rivers, below I examine each of the four dimensions proposed here.

Defining economic justice: Economic justice refers to the principle of distributive fairness widely discussed within environmental justice literature and political philosophical traditions. In environmental movement discourses, economic justice is defined as a pursuit of distributive equity of both environmental goods and bads. While distribution of environmental goods refers to dissemination of environmental resources for all communities irrespective of their social location, environmental inequities should not be shouldered alone by disenfranchised groups (Bryant & Mohai 1992). Particular attention has been paid to studies of disproportionate burden of toxic contamination among the poor communities and communities of color in the United States (Bullard 1994; Freeman 1972; United Church of Christ Report 1987), historical and institutional nature of unequal distribution, inception of environmental justice movements (Pellow 2004; Taylor 2000). Recent research on unequal consequences of global warming, food justice struggles, and global waste transfers have significantly internalized the studies of distributive/economic concerns in global environmental movements (Checker, 2010, Dorsey, 2007, Roberts & Park 2009, Stallworthy 2009).

This notion of distributive fairness has been debated eloquently in political philosophy. Fairness in justice theory, John Rawls argues, is mainly concerned with the need to pursue equality in distribution of all social values (Rawls 1971, 1995). For Stephen, Lukes, and Amartya Sen, opportunities for just distribution are realized when we create equitable political arrangements, either through building capabilities and/or through initiating a social order devoted to principle of justice and equality. Environmental justice scholars join political philosophers in these assertions. The pursuit of principles of fairness and freedom can only occur by building people's capabilities, which in turn requires a social order based on principles of dialogue and democracy. Equitable distribution of material resources includes not only the distribution of resources (a social scientific concern), but also the capability to pursue these resources within a framework of dialogue and democracy (a political philosophical concern) (Banerjee & Bell 2007). In other words, fairness in both distribution and sociopolitical conditions are essential components of a justice theory.

Fairness as an element of justice theory refers to distributive equity of both environmental goods and bads. The fairness principle also underscores the need for accessibility and accountability of environmental knowledge for communities, especially in the aftermath of an environmental disaster. It refers to the need to disseminate "expert"—scientized knowledge and practices for greater participation by affected communities. Thus for example, Hurricane Katrina brought to light the consistent underinvestment in disaster prevention and preparedness for marginalized communities. It also highlighted the complicity of state and federal agencies' complicity in the current disaster and the state of Louisiana's abysmal record of non-protection of its residents (Bullard & Wright 2009: 117). This inattention to the fairness principle is particularly evident in the inattention to arsenic contamination discovered in schoolyards in post-Katrina New Orleans due the a lack of clarity about hazard levels of arsenic and whether it was present before disaster struck (Godsil et al. 2009: 123). Similar concerns about lack of fairness in creating scientific standards for hazardous wastes can be perceived in the TVA ash spill disaster in Kingston, TN with EPA's controversial standards for storage of coal ash.[6] The fairness principle thus addresses key questions about material equity building and distribution of resources, wastes, knowledge for a just social order.

Defining cultural justice: Conflicts over environmental justice are neither solely a material concern nor do they rest solely on capability formation. Instead, conflicts over environmental resources are often expressed through non-material/cultural lenses. Demands for the "recognition of differences" finds expression in Nancy Fraser's writings on theories of justice, but have been examined recently in Schlosberg's research on how non-material identities are essential components of environmental conflicts (Fraser 2009; Schlosberg 2007). Others like Kosek (2004) and Kuletz (1999) examine the cultural politics of memory and longing in loss of land and identity in northern New Mexico and the manufactured invisibility of Native Paiutes in the deserts of California, respectively. Notwithstanding these and other exceptional contributions, environmental justice research remains predominantly distributive in nature.

Yet, conflicts over environmental injustices rarely end with a redistribution of resources. Instead, they frequently call for rehabilitation of communities in the aftermath of a disaster, either by moving to a new place or by remaking their homes. These initiatives occur against the backdrop of a place that is intimately connected to the identities of the displaced community. Thus for example, the oral histories gathered in post-Katrina New Orleans frequently refer to the cultural loss of communities who must now redefine their histories, recuperate from the loss of their memories, homes, and places as they deal with grave environmental health violations. Environmental justice conflicts after all occur in place and must be incorporated within the discourse on justice. As political philosophers like Iris Young states, distributive models of social justice tend to calculate costs and benefits in determining material inequities and when cultural contexts enter the framework (as they do occasionally) they are viewed in static material terms (Young 1990; Warren 1999).

Environmental conflicts occur *in place* and often result from violation or destruction of place and by extension cultural identity of a group. Cultural justice incorporates two specific justice concerns. First, it allows us to explore the importance of place-based narratives in shaping environmental conflicts. Cultural narratives of justice include studies of discourses, heritage, community identity, and place and empower communities to organize resistance movements. As the story of Land Between the Rivers suggests, the environmental movement was precipitated by displacement and loss of cultural heritage and place. For the community their informal environmental initiatives were deeply influenced by their close relationship with their place. As the movement evolved and the community improvised and chose economic, institution, and political resistance strategies, loss of place and cultural heritage retains its importance in the community's discourses.

Second, cultural justice is essentially a material and political struggle for equality and must be examined as a dynamic resistance to loss of identity and agency. Conflicts over distribution of environmental resources are not immune to ideological, moral, or political claims. The elimination of people from their place changes the ways in which resources are perceived, valued, allocated, and used. For the residents of Land Between the Rivers, the cultural injustice was intensely political. The cultural stereotyping that preceded the displacement, the deeply flawed reconstructions of the history of displacement, and the non-recognition of the community's rights to their cultural identity were all political acts. For the former residents, the lack of acknowledgment of human role in co-creating nature is a political statement involving calls to recognize identities and histories.

Similar references can be found in scholarly research on justice. Fraser examines how communities influence the plural perspectives of justice (Fraser 2009). Others like Schlosberg explore the role of cultural differences in construction of environmental conflicts (Schlosberg 2007). Environmental geographers like Kosek (2004) and Kuletz (1999) examine how collective loss and memory emerge from "manufactured invisibilities marginalized groups in New Mexico and California, respectively. The importance of plural approaches to justice is also found in political philosophical discourses of Young (1990) who explore cultural contexts in justice narratives. Recent research on disaster studies adopts a similar approach. Allen (2011) refers to displacement amid disaster as communities in New Orleans seek to rebuild their homes, rewrite their histories, and recuperate from the loss of their heritage and cultural identity.

Cultural narratives of justice in environmental movements offer unique opportunities to learn how movements emerge and coalesce over time. First, conceptualizing the cultural dynamics in environmental justice conflicts allows us to examine the hidden understories and community narratives that inform these conflicts. The environmental narratives about recent BP oil spill along the Gulf coast examined the political of oil in shaping the disaster, the cleanup and its organizational dynamics, the political economy of communities, and the environmental health impact of affected community.

Second, just like distributive models of environmental justice, cultural concerns are not necessarily apolitical in nature. Conflicts over distribution of environmental resources are not immune to ideology, idealism, or moral judgment—all of which have political components. The elimination of people from their place changes the ways in which resources are perceived, valued, allocated, and used. Human agency (investment of labor) in creating nature (place) imparts rights of ownership or access (Smith 2008). In other words, lack of acknowledgment of human role in co-creating nature *is* a political statement involving calls to recognize identities and histories.

Further, examining the gamut of cultural narratives also opens up possibilities for studying the use of cultural tools (often adopted by affected communities) in resisting environmental decisions. Thus for example, an encultured environmental justice approach highlights a wide spectrum of both traditional and non-traditional forms of resistance in the course of an environmental conflict. It allows us to understand how non-participation is a form of tacit resistance to dominant portrayal of environmental conflicts. Thus acts of refusal to accept compensation for environmental violations by community members can be viewed as passively expressed cultural resistance; further, active forms of resistance include building coalitions, presenting stories to the media outlets, community meetings, and engagements; and finally, place-based contestations also mount extensive challenges to dominant environmental narratives, which are often place-less and global (as is evident in the successful challenges of environmental justice movements to the "wilderness preservation" approaches of traditional environmentalism). Adoption of these widely understudied cultural protest tools will allow environmental justice movement key and important tools to document how protest movements evolved from collective attachment to place and concerns about the loss of this cultural link.

Defining processual justice: In addition to cultural and economic dimensions, environmental justice also includes the analyses of institutional arrangement and democratic participation, or *processual justice*. This refers to a social arrangement in which equitable participation, open deliberation, shared environmental knowledge creation, and institutional transparency. Environmental justice scholars have discussed the role of institutions in shaping justice (Pellow 2005). Others have examined the role of political participation in fostering institutional justice (Scholsberg 2007, Ukridi & Walter 2011). Further, researchers in Science and technology studies provide normative institutional frameworks to explore how democratization of scientific "expert" knowledge promotes environmental justice. Jasanoff and Wynne (1998) propose the need for a democratic platform for effective collaboration between citizens and experts on scientific (and environmental concerns). Others like Ottinger (2010) calls for the need for better conceptualization of standards of legitimation for citizen versus scientific knowledge claims while Frickel (2007) examines the inequities in the construction and dissemination of environmental knowledge and the need for more equitable and transparent institutional framework.

For example, in the community of Land Between the Rivers in western Kentucky, participation in environmental decision-making remains deeply linked to their pursuit of cultural heritage and environmental justice. The residents' demand for processual justice emerged from their struggle against cultural stereotyping, non-recognition of their displacement history, commercialization of their land, and unilateral decision-making on environmental policies in Land Between the Rivers. The recent acknowledgment of the former residents as a consulting party to heritage management was a positive step in this direction. Similar claims to institutional participation have emerged in struggles against food injustices, climate change discourses, and community land preservation movements around the world. For example, in his studies of poverty and famine welfare economist Amartya Sen distinguishes between food availability and food entitlement (Sen 1981). In his words an individual's entitlement set consists of a range of goods and services that she can acquire by converting her assets and resources (labor power) through fair exchange of entitlements. A famine results from a collective inability to engage in these exchanges. Individuals face starvation if their full entitlement set does not provide them with adequate food for subsistence. Sen's entitlement approach is particularly relevant to environmental justice conflicts. His approach allows us to explore how differently positioned social actors command environmental goods and services shaping community well-being.

Sen's notion of environmental entitlement allows us to examine how strong and effective institutions can not only facilitate simple access to environmental resources like food, but (more importantly) encompasses the ability to make effective use of these resources (Sen 1981). More specifically, the idea of an entitlement set allows affected communities to examine possibilities for meaningful participation in decisions about sharing environmental resources, goods, and knowledge. As environmental justice struggles occur globally, pursuit of processual justice with its emphasis on equitable participation and democratization of environmental knowledge will gain further prominence in movement and academic discourses.

This notion of equity and democratic processes finds expression in political justice as well. For example Amartya Sen states that the principles of fairness remain unattainable unless accompanied by the freedom to choose. The notion of choice as Sen defines it is essentially non-utilitarian and is the essence of his entitlement approach. In examining the immediate factors that lead to famines, Sen distinguishes between food availability and food entitlement (Sen 1984). In his words an individual's entitlement set consists of a range of goods and services that she can acquire by converting her assets and resources (labor power) through fair exchange of entitlements. A famine results from a collective inability to engage in these exchanges. Individuals face starvation if their full entitlement set does not provide them with adequate food for subsistence. The entitlement approach is primarily useful in shifting conceptual focus from a Malthusian analysis of overpopulation as the primary cause of poverty to an examination of food insecurity and inability of groups to access resources.

Sen's entitlement approach is particularly relevant to examining how the pursuit of processual justice remains the cornerstone of environmental justice mobilization. His approach allows us to explore how differently positioned social actors command environmental goods and services shaping community well-being. While he views entitlement primarily as a material/distributive question, his notion of capability allows us to examine how groups negotiate their positions vis-à-vis available goods. Sen's framework thus allows us to avoid the distributive predicament of the characteristics of available goods and underscores the need to value political/institutional arrangements that enable communities to access these goods. Thus, the notion of environmental entitlement refers not only to the simple access to environmental resources, but (more importantly) encompasses the ability to make effective use of these resources. More specifically, the idea of an entitlement set allows affected communities to examine possibilities for meaningful participation in decisions about sharing environmental resources, goods, and knowledge. In turn this helps in the capability formation for communities as they pursue the goals of fairness and allows space for agency.

Defining Political Justice: Resistance narratives, James Scott argues, are often tacit and covert acts of protests, particularly evident when institutional justice opportunities are either limited or largely ineffective (Scott 1985). In *Weapons of the Weak*, Scott studies the ordinary, everyday forms of resistance that peasants engage in, actions that are covert but calculated. Beneath the peasants' apparent compliance, he says, are layers of ideological and symbolic resistance. Scott separates these forms of resistance from the "genuine" ones, resistances that involve organized and systematic uprisings, much as Marx favored. Political Justice encompasses both these covert and overt forms of environmental activism. The tactics adopted for seeking attention to one's movement goals may involve a series of strategic choices. For example, movement groups may seek to pursue strategies of practical politics or "politicking" to identify and build coalitions with groups that share similar political goals.

These politicking strategies include building campaigns for sharing political messages with the public, lobbying interest groups to publicize grievances, build coalitions with similar groups, and demand effective policy solutions. One example of politicking can be observed in the course of the women's suffrage movement in the United States in the early twentieth century when the "National Women's Party (NWP) engaged in politicking through parades, pageants, street speaking, civil disobedience demonstrations, and mass meetings in order to aggressively seek legitimacy and support" (Banerjee 2014: 6)

Further, movement groups may also seek success by appropriating ideas and narratives of social justice from "established" movements. These appropriations can occur as movement groups draw conceptual and philosophical parallels with successful social justice movements. Such linkages allow aspiring movement groups to frame their concerns as an element of larger struggle thereby gaining legitimacy and support in the public sphere. For example, in the case of the women's suffrage movement, the NWP drew inspiration from a variety of established sources, including the British Suffrage movement, labor movement, temperance and antislavery initiatives, and the early American Suffragist struggles[7]. Similarly, the environmental justice movement garnered public support by linking its message of environmental equity with the broader social justice discourses of the civil rights and labor movements (Taylor 2000).

This is also evident in the environmental movement in Land Between the Rivers. For example, the community's insistence on naming the place as

Land Between the Rivers is an act of covert resistance to the federal renaming of the refuge as Land Between the Lakes. For the former residents, the act of naming was an opportunity to hold on to bits of history that defined their connection to the place. Similar acts of political justice include refusal to accept compensation from TVA, rebuilding the local church, spring cleaning, and preservation of the cemeteries. Finally, as discussed earlier, recent conflicts with the Forest Service has led to parallel community initiatives for cultural tourism, development and implementation of new projects in the area, and lobbying for signatory status in environmental and cultural decision-making. Pursuit of political justice incorporates elements of cultural, economic, and institutional justice, yet allows affected communities to engage in alternative modes of contestations over environmental rights.

The acts of politicking for justice assume greater importance as environmental movements gather global momentum. The growing internationalization of environmental justice movements calls for greater attention to the specific need to conceptualize the role of the political structure in either facilitating or hindering environmental negotiations. Thus for example, the new geopolitical and economic realities in India are reflected in the state's invitation to Union Carbide (the corporation involved in the Bhopal Gas Tragedy of 1984) to reenter the market. The emergence of India as a newly minted economic powerhouse is accompanied by a steadily eroding space for contestations over human rights and environmental violations. Yet, the availability of a democratic platform (an imperfect one at best) has contributed to the possibility for mounting counter campaigns or sustaining the civil groups' fight for justice for Bhopal victims and their families. While the political contexts are vastly different, the state strategies adopted to mount counter pressures intimately shape environmental negotiations. Hence, a justice theory needs to critically identify the nature of the democratic framework, the role of civil society organizations, and the political strategies adopted by the state in influencing environmental justice conflicts and their resolution.

Concluding Thoughts

Recent years has witnessed the emergence of a critical mass of scholarship on theory building in environmental justice. This chapter builds on these traditions of scholarships and proposes a framework that links the sub-discipline with political philosophical approaches to the question of justice. The proposed justice framework addresses a broader range of inequalities, including questions of the fair distribution of environmental goods, creating capacity for access to these goods, a cultural/non-material approach to environmental justice, and a need to examine the political/democratic arrangement in environmental decision-making. Revisiting justice using a rights-based multilayered plural approach has significant benefits.

First, it allows us to conceptualize environmental justice as more than a question of distribution. Instead, a justice theory, seen through the lens of human rights and tempered with a political philosophical understanding of capabilities, highlights the intellectual possibilities that lie nascent within environmental justice theory. Thus for example, the environmental rights, like right to a pollution-free environment, right to a healthy life, and so on, are most achievable only when the relevant capabilities to function are present (Sen 1992). Second, an environmental rights approach also includes community concerns with cultural property rights—both material (artifacts) and non-material (stories, narratives, oral histories, music, and so on). Finally, this framework allows us to account for the characteristic of the institutional framework for the successful implementation of these rights. We can only call a society just when it has created effective conditions for achieving these capabilities. Hence drawing linkages between rights and environmental justice through capability building will provide a more effective framework for conceptualizing agency within environmental justice theory (Banerjee 2006).

Two decades ago, environmental justice research and activism engendered a radical shift in our collective consciousness about how environments often emerge as sites of inequality and violence. It was a global wakeup call, one that allowed us to transcend the debates on conservation and explore how socially embedded inequities inform environmental distributions. Environmental justice stands at a new crossroads with an increasingly diverse and overlapping range of conflicts that include material inequities, human rights violations, and cultural losses. Our challenge lies in making room for this diversity without losing the conceptual strengths of the framework. The four-dimensional framework proposes a conceptual approach to document the fluid, multilayered, and plural nature of the concept of "justice" in environmental justice. It allows us to examine how environmental justice discourses evolve over time and adopts multiple narratives as it struggles to gain legitimacy and support. Together, the four justice strategies help us understand and theorize the fluid nature of environmental justice movements.

The next section of this book offers a personal narrative of the author as the researcher in the ethnographic study in Land Between the Rivers. My research was conducted over several months and involved deep immersion in the field as I learned the historical context and the contemporary challenges of the environmental movement in Land Between the Rivers. I build on Georg Simmel's essay "The Stranger" to explore how ethnographies shape the researcher's evolving identities and various roles that she adopts in the course of the research.

NOTES

1. Agyeman, J., R. D. Bullard, & B. Evans. 2002. "Exploring the Nexus: Bringing Together Sustainability, Environmental Justice and Equity." Space & Polity. 6(1): 77–90.
2. Bryant, B. & P. Mohai. (eds.). 1992. Race and the Incidence of Environmental Hazards: A Time for Discourse. Boulder, Co: Westview.
3. Pellow, D. N. 2004. "The Politics of Illegal Dumping: An Environmental Justice Framework." Qualitative Sociology. 27(4): 511–525.
4. Taylor, D. 2000. "The Rise of the Environmental Justice Paradigm: Injustice Framing and the Social Construction of Environmental Discourses." American Behavioral Scientist. 43 (4): 508–580.
5. Holifield, R, M. Porter, & Walker. 2009. "Spaces of Environmental Justice: Frameworks for Critical Engagement." Antipode. 41(4): 591–612.
6. Recent EPA hearings held in Knoxville TN open up the belated possibility for establishing better guidelines for managing fly ash.
7. http://memory.loc.gov/ammem/collections/suffrage/nwp/tactics.pdf

Part III

RESEARCHER STATEMENT

Chapter 2

The Narrative Moment

Informing Environmental Claims through Ethnographic Analysis

I vividly remember the day I visited a workshop on place with students and faculty of a local university at my research site in Land Between the Rivers in western Kentucky. I arrived just as the late afternoon sun was losing some of its brilliance and promptly joined a group of students and faculty examining one of the cemeteries of Land Between the Rivers. We were discussing what a sense of place meant as we walked among the headstones. I was, as is usual in most of the gatherings that I went to, the only non-white person in the group. This was not uncommon, but as I joined the group I felt out of place as never before. I felt, as many researchers feel while working in "unfamiliar" field settings, uneasy in my skin.

Afterwards, I wondered if this feeling of uneasiness is what Simmel referred to when he was discussing the positionality of the "stranger" (Simmel 1950). In fact, this is particularly relevant to my work in Land Between the Rivers where I was conducting a participatory ethnographic field study over a period of several months. While I entered the field as a stranger/outsider to the community, my subsequent experiences added multiple layers of meanings to my interactions and engagement with the community. In this sense, what began as ethnographer-as-stranger evolved into a greater embeddedness and closer engagement with the local community. This chapter seeks to explore the specific place and role of the ethnographer in her field and its impact on researcher and data collection.

Analysis of the role of the "ethnographer-as-stranger" is by no means a novel approach. Sociologists have long been interested in examining the positionality of the researcher in her field and the impact of the interactions of the researcher with the researched in the narrative process. Hammersley and Atkinson (1995) have examined "arrival" stories of researchers and their shifting roles in the field. Others have explored the fragmented self

(Kondo, 1990), rewriting identities (Mcintyre 1993), resistance and negotiation of identities (Blackwood 1995), marginality in the field (Hammersley & Atkinson 1995), ethnography as immersion (Peshkin 1985). Still others have explored the shifts in positionality from that of a marginal outsider to an engaged participant in a cultural setting with challenging, often contradictory roles (Blackwood 1995, Hendry 1992).

Yet as Amanda Coffey (1999) succinctly pointed out, the notion of the stranger-ethnographer, in conventional writings, is misconceived at best. For the ethnographer's journey is a complex and delicate process of negotiation, one that does not follow a straight and narrow path from ignorance to enlightenment. Instead Coffey invites us to understand the ethnographic immersion in the field that captures the difficult and often contradictory location of the researcher as one who is "both marginal to and fundamentally engaged in a cultural setting" (Coffey, 1999: 32). Coffey's call for a complex analysis of locationality in ethnographic writings emphasizes the importance of our bodies in the research process. This chapter applies this notion of complex locationalities to discuss the importance of the ethnographer's multiple identities and shifting and evolving affiliations. Following Coffey, I argue that the ethnographic self is essentially an embodied identity engaged in a journey. I explore this embodied journey by using the metaphor of the skin.

The metaphor of skin is closely linked to the notion of the stranger discussed in Georg Simmel's writings. Simmel's stranger, as he aptly conceives her/him, is the eternal outsider—one who lives in her/his skin but is also an outsider to it. As he points out,

> To be a stranger is naturally a very positive relation; it is a specific form of interaction. The inhabitants of Sirius are not really strangers to us, at least not in any logically relevant sense: they do not exist for us at all; they are beyond far and near. The stranger, like the poor and like sundry "inner enemies," is an element of the group itself. His position as a full-fledged member involves both being outside it and confronting it. (Simmel, 1950 [1908]: 402)

The ethnographer is similarly confronted by the conflicting, yet a "very positive relation" with the other—as both a member of the community and an outsider. The stranger's "double consciousness" emerging from conflicting identities and complementary locations is reminiscent of DuBois's reference to the duality of identities—frequently observed among minorities and immigrants (Du Bois 1999 [1903]). I wish to suggest that such duality (and multiplicities for that matter) is similar to the experiences of the ethnographer's journey, that it segues into the narrator's being, and that it explores the juxtaposed identities which inform the research process. Like Simmel's stranger then, the ethnographer remains an outsider to her field (Simmel 1950

[1908]). Yet, there are differences as well. Unlike Simmel's stranger, who, in his words, "comes today and stays tomorrow," and hence is a full-fledged member of the community in a temporal sense, the stranger-ethnographer "comes today and leaves tomorrow" and wears her many identities, as a temporary member of the community, researcher, and collaborator.

For the researcher, these multiple associations are most aptly explained using the metaphor of skin. First, underscoring the embodied nature of the ethnographic self helps us in conceptualizing the complex negotiations between the researcher and her field. Second, the ethnographer's journey through her field resembles our individual experiences in and out of our skin as we attempt to reconcile our varied and often conflicting identities and values. I introduce the notion of the skin as a multilayered and extendable identity that we periodically adopt and discard as we conduct our researcher. The metaphor of the skin refers to our racial, class, gender, academic locations that strongly shape our interactions with our research partners in the community and provides us with an understanding of the continuous negotiations of identities in the process of story-telling. I argue that our identities, like our metaphorical skin, are extendable and transform themselves as we explore the contours of life. Each individual experience determine the path that our identities take, a path that sometimes travels through our skin and at other moments expands our metaphorical skin to include both our personal and professional locations.

In the final section of the chapter, I amplify my argument about the stranger-ethnographer's journey by narrating a story of a small forest community called Land Between the Rivers in western Kentucky. In narrating the story, I look at my shifting locations as I attempted to negotiate my multiple identities in the field—as a woman of color, an academic, an upper-middle-class scholar of non-American nationality working in a predominantly white community in the rural South in the United States.

RESEARCHING THE ETHNOGRAPHER'S LOCATION: THEORETICAL PERSPECTIVES

Ethnographers have often discussed challenges that researchers experience while out in the field. Early social anthropologists like Malinowski set the path for an ethnographer's observations—on his/her experiences working with the "natives" trying to "grasp the native's point of view, his relation to life, to realize his vision of his world" (Malinowski 1961: 25, emphasis in the original). Yet, the vision of the world that Malinowski was referring to was essentially a vision from afar, his own interpretations of the "native's" culture. Malinowski's writings were primarily objectivist in nature, with no attempt to explore the impact of the ethnographer's subjectivities on the research process.

In subsequent writings, questions of subjective experiences and objective analysis have emerged as a central dilemma in sociological research. Recent years have witnessed considerable emphasis on the study of ethnographic research as a participatory enterprise. In particular, sociologists have examined the role of ethnographic experiences in the research process, including the study of the ethnographer's mind as the transitional space for analyzing the observed data (Churchill 2005), the role of the collaborative ethnographer (Duneier 1999, Gans 1962, Gottlieb & Graham 1992, Gupta & Ferguson 1997, Whyte 1981), phenomenological examination of ethnographer's position (Davidman 1997, DeVault 1997), and ethnography as a personal narrative (Cook 2005, Coffey1999, Pierce 2003).

Similarly, auto-ethnographers have consistently emphasized the location of the ethnographer as a "full member in the research group or setting" (Atkinson 2006). In fact, a collection of articles published in August 2006 in the *Journal of Contemporary Ethnography* addresses the role of the ethnographer as an emotive and analytic member of her research setting.[1] Other scholars have explored the benefits of conducting research as an engaged experience with participants as stakeholders. Scholars within the participatory research tradition have often argued that participatory research ultimately attempts to positively contribute to building relationships of trust and legitimacy through what Chambers calls one of the most important philosophical tenets of participatory research: sharing knowledge and experience (Chambers 1994). Still others point out participatory research tradition's commitment to display social realities as they are lived, experienced, and understood by humans (Katz 1997). Thus, recent ethnographic research has clearly diverged from the objectivist understanding of the research subject as seen in Malinowski's writings and has made significant contribution to the exploration of the researcher's location in the research process.

This chapter builds on these conceptual tools to explore the unique positionality of the ethnographer whose journey begins as a stranger and evolves to that of an invested collaborator, actively engaged in shaping the narrative and being shaped by it in turn. More specifically, I explore the impact of multiple identities of the ethnographer in her story-telling initiatives. My ethnographer enters her milieu as a stranger and continuously travels through her skin (both in and out of her body) and emerges from the process with a more situated understanding of the stories that she narrates.

The ethnographer's journey makes her a stranger in her skin as she enters her research site. Following Simmel's excellent essay on the stranger, I argue that the position of the ethnographer is partly that of a stranger in a group who "has not belonged to it from the beginning, that he imports qualities into it, which do not and cannot stem from the group itself" (Simmel 1950: 402). Yet, the ethnographer is more than a "potential wanderer, one who comes today but stays tomorrow" (Simmel 1950: 402). Instead, she "comes today and leaves tomorrow" and her position is shaped by the two-ness of her existence as "one who lives in two worlds" (Park 1937). An ethnographer battles with her double consciousness as one who is "always looking at one's self through the eyes of others" (Du Bois 1999). An ethnographer then is a sojourner in her own skin, traveling in and out of it as she negotiates her positionality in her research milieu.

Second, I examine this experience of sojourning in the course of my own research experience in a small community called Between the Rivers located in the western fringes of Kentucky. I examine how my own social location, as an upper-middle-class academic, a woman, a person of color, and a foreign national shaped my research in Between the Rivers. I explore how these multiple locations have enriched my research, while trying to keep in mind what my class/ethnic/gender/race locations may have prevented me from seeing, just as they gave me some vision that I might not otherwise have had. Finally, I end with a brief discussion of possibilities for more research on the ethnographer as the sojourner.

Sojourning in my Skin: Researcher Positionality in Qualitative Sociology

In the book *The Color Purple*, the protagonist Celie tells us the trials of being the "lowest of the low, put upon both as a woman and a woman of color" (Walker 1982). She grows as a woman as the story evolves, charting the tricky bends of identity as she sojourns in her skin both as a woman and a person of color. Her sojourns through her skin provide her not only with an understanding of her race, but also of her gender. Celie's sojourns provide her with a unique understanding of social locations that shape her various subjectivities. Sojourning in and out of one's skin then involves more than experiencing the color of one's own skin. Instead, it involves negotiating multiple identities along lines of race, class, gender, ethnicity, and social location.

Such negotiations of identities and social place are not uncommon in sociological writings. Sociological writings in the early- to mid-twentieth century witnessed the enormous popularity of Georg Simmel's concept of "the stranger" (1908), so much so that an entire gamut of literature on the "sociology of the stranger" was produced (Catton 1964, Nisbet 1966, Nash & Wolfe 1957). Simmel's insights on the stranger were exceptional in that his stranger was a person who, while a member of a community in a spatial sense, was still an outsider in a social sense. In other words, Simmel pointed out that a person may belong to a group, but not of it (Wood 1934). Because of her/his position within a community, the stranger for Simmel is a more objective member of the group who enjoys greater freedom and a more incisive understanding of the group. The arrival of the stranger poses a challenge to the existing set of social relationships in the group and consequently impacts the social organization, interaction patterns, and community relationships.

Simmel's notion of the stranger is useful in examining the role of the ethnographer in the field. The ethnographer is a stranger to her social world, yet her engagement with the community needs to be intimate given the nature of ethnographic research. Just as the stranger's interactions with the community change as she/he "moves from outside the society toward the inside," the ethnographer becomes an invested collaborator in the research process as her narrative unfolds (Malouf 1978). Further, given her outsider status in the community, the ethnographer is assumed to be more objective in dealing with emotive community concerns. Like Simmel's stranger who as an outsider, "is not radically committed to the unique ingredients and peculiar tendencies of the group, and therefore approaches them with the specific attitude of 'objectivity,'" the ethnographer's "expert" location provides a sense of objectivity to the research process (Simmel 1950 [1908]). Objectivity, however, as Simmel argues does not imply passivity and detachment. As he suggests, "it is a particular structure composed of distance and nearness, indifference and involvement" that informs the researcher's objectivity and which "occasionally receives the most surprising openness--confidences which sometimes have the character of a confessional and which would be carefully withheld from a more closely related person" (Simmel 1950 [1908]).

The ethnographer's objectivity is not a passive distancing emerging from non-participation but one that is brimming with activity and interactions that is both participatory and involved. Yet, there are startling differences in the strangeness of Simmel's stranger and the ethnographer. Unlike the stranger, the ethnographer loses some of the objectivity as her own locations and identities pull her further into the community activities as she transforms into an invested collaborator. While Simmel's concept expands on the impact of the stranger on the social milieu of the community where she/he lives, it does not clearly demonstrate how the stranger's own positionality changes and transforms the stranger. Simmel's stranger possesses qualities of "nearness and detachment, concern and indifference" and whose "relative detachment frees him from the self-consciousness, the concern for status, and the divided loyalties ..." (Stonequist 1937: 177–78). On the other hand, the ethnographer is emotionally invested in the research process, and her subjective understandings of community narratives shape the relationship between the community and the stranger. The ethnographer is one who, having taken on elements of the dominant culture, is unable to return unchanged to her/his original group.

I argue that the ethnographer's location then hints at a certain amount of openness of identity and can be explored as a sojourner in her own skin. The Oxford English Dictionary defines "sojourner" as "one who sojourns," "a temporary resident," "a visitor," and "boarder living in a house, school, or college, for the purpose of receiving instruction." The ethnographer, however, is not only a sojourner in a spatial sense, as who "comes today and leaves tomorrow." Instead, she sojourns in her own skin as she makes sense of her own subjectivities and the subjectivities of her research participants. While a temporary resident in the community, the ethnographer evolves along with the research process and her shifting perceptions shape the research narrative. In his classic study of Street Corner Society William Whyte examines the role of the researcher in the research process. The researcher, he argues, "has a role to play, and he has his own personality needs that must be met in some degree if he is to function successfully. … A real explanation, then, of how the research was done necessarily involves a rather personal account of how the researcher lived during the period of study" (Whyte 1981: 279). In the role of the sojourner, the ethnographer is able to constantly examine her own positionality that simultaneously allows her glimpses of the community while barring access to others.

In the section below, I explore these seeming contradictions in examining my own stay in a community called the Land Between the Rivers. Located in the borders of western Kentucky, the forested and hilly landscapes of the region have led to the construction of a community identity that is clearly defined and closely knit. The community of approximately 1000 families had settled in the forested land in the late eighteenth century and has had a long association and history in the region. However, with the economic developments of the 1930s under the Tennessee Valley Authority (TVA) the residents were removed, often forcibly, in five phases over a period of four decades for dam-building in the two rivers that surround Between the Rivers. Finally, in the 1960s with the removal of the last residents, the area was declared a federal property and earmarked for conservation and recreation and renamed Land Between the Lakes (LBL).

Residents of the area have been actively engaged in resisting displacement from the region. In recent years, resistance has been more focused on striving against commercialization of land, and focusing on preservation of cultural property and community identity. The issues have remained politically charged with conflicts over access rights, community diversity, and cultural legitimacy. When I visited Between the Rivers in late October 2003, Between the Rivers has been transferred from the TVA to the Forest Service and the former residents were engaged in attempts to gain legitimacy for their cultural identity. The lines were clearly delineated and the struggle was well-developed as I set foot in the region leaving me with a critical question—how can I contribute in a useful way and work effectively with the groups involved for a resolution of the crisis?

I was at first unsure about what my contribution can be to the process. I was after all not a cultural lawyer who could be an effective element in pursuing the struggle for cultural identity. Nor was I an anthropologist, with special skills necessary for identifying cultural resources and artifacts instrumental in gaining cultural legitimacy. I wasn't even an experienced social movement activist whose expertise in skillful negotiation with state agencies might help in the struggle for identity. Also, it seemed to me that the concerns of the community were already clearly developed and local activists were particularly adept at articulating their grievances and concerns with the state preservation agencies.

I was, to own the truth, seriously concerned about my own inadequacies to be an effective contributor to the struggle for identity in Between the Rivers. Unable to bear my misgivings, I asked my community partner his opinion about the best possible way I can help their struggle. Fortunately, he was very clear in his perceptions of what might be beneficial for the community and pointed out that one of the important ways in which I can help is by providing a more objective understanding of the concerns to which the community residents, owing to their history, are emotionally connected to. In other words, he viewed my position as an outsider as instrumental in providing some objectivity to their struggle for legitimacy. I did find this reassuring for here was a situation where my social location as an outsider was an effective tool for the community. As I discovered in time, the seeming objectivity was continuously challenged by my subjective self and its multiple positionalities. My social location in Between the Rivers shifted constantly as I was called upon sometimes to work in my skin (in a race/class/gender sense) and sometimes outside (as an objective outsider). At other times, my subjectivities changed as the community boundaries extended or contracted like the skin of my body, as I became a participant to some, while an outsider to others. Thus, I straddled multiple identities in the course of my research, which while a mixed blessing, has been instrumental in my understanding of the emotive effect of displacement on individual and community psyche. I discuss these issues in the section below.

Identity Construction in the Ethnographic Field

Every time I am asked this question, I lapse into long periods of self-reflection. A simple answer of course would be to point out that I am a citizen of India residing here in the United States for graduate studies. However, while this is a correct response, it does not account for all of me. I am a Bengali and an Indian whose parents were first-generation migrants from Bangladesh displaced in the aftermath of a violent partition and who have lived outside Bengal and India for a considerable amount of time and hence is somewhat of a self-identified global citizen. Yet, these self-assigned identities only partially explain my own positionality. Identities, especially when attributed by the "man of the street" are often based on color of one's skin. Brown, black, white, yellow (may be blue and purple as well, if we consider "aliens" and attribute colors to them). Identities are also based on one's cultural location and social tastes. Based on our cultural, political, environmental leanings, we are also green, blue, red individuals. Together, these various identities shape our sense of self through processes of negotiation and self-reflection. Yet, none remain unchanged as we negotiate our positionalities depending on our own lived experiences. In his recent book on Identity and Violence (2006), Amartya Sen observes these negotiations and contestations of our multiple identities during times of stress and violence, problems symptomatic of our current times. According to Sen:

> In our normal lives, we see ourselves as members of a variety of groups ... the same person can be, without an contradiction, an American citizen, of Caribbean origin, with African ancestry, a Christian, a liberal, a woman, a vegetarian, a long-distance runner, a historian, a schoolteacher, a novelist, a feminist, a heterosexual, a believer in gay and lesbian rights, a theater lover, an environmental activist, a tennis fan, a jazz musician, and someone who is deeply committed to the view that there are intelligent beings in outer space with whom it is extremely urgent to talk (preferably in English). ... None of them can be take to be the person's only identity. ... Given our inescapably plural identities, we have to decide on the relative importance of our different associations and affiliations in any particular context. (Sen 2006: xiii)

Sen's reference to the dilemma of multiple identities in times of inordinate stress (e.g., a violent riot) resonated with me as I embarked on my research in Between the Rivers. Unlike, Sen's protagonists I was not facing violent social outbursts, yet I was engaged in similar negotiations in order to build trust and create relationships within the community. Such negotiations involved transcending boundaries that our social locations impose on us and which shape the participatory research process.

In a recent paper, Bagby and Kusel point out that the key dimensions of participatory research include "developing trust with communities, building effective partnerships, engaging with cultural concerns and contributing to a greater dialogue and empowerment that provide long-term benefits to the community" (Bagby & Kusel 2002). As I entered my field, my primary concern was to cultivate these relationships of trust and mutual engagement. This was not an easy task given my own social location in the milieu of Between the Rivers. I was, after all, an upper-middle-class reasonably privileged academic, and a non-white foreigner trying to find a foothold in an unfamiliar country. As a woman of color in a post-9/11 America, I was attempting to work in a community which is predominantly white and situated in a rural setting in the U.S. South. I was then, in certain ways an anomaly in the world of my subjects trying to gain an understanding of the concerns of the community of Between the Rivers with whom I have little in common (Wells 1996).

Yet, in other ways, I was not so much of an outsider. Displacements, after all, are problems that affect millions in the world. My own experiences in India with an ongoing conflict over dam-building and displacement of millions in the Narmada valley have provided me with a general understanding of some of the concerns with place, cultural heritage, and community identity. Concerns with state construction of environment in place-less alienating terms are familiar stories in the developing world. Further at the community level, I was fortunate enough to gain unstinting support from a local community activist, who luckily for me was also a fellow sociologist! As he guided me through the complicated community concerns, I became over time less of an outsider (at least in my own mind, and probably to some extent in the minds of those I worked with). My frequent interactions with him were instrumental in gaining a clearer understanding of not only the history of Between the Rivers but of some of the historical narratives of the U.S. South. Additionally, the community itself was also a willing and enthusiastic participant in patiently explaining the intricacies of the debate over the history of dam-building in Between the Rivers.

Yet, our interactions began on a distinctly uncomfortable note during one of the early focus groups with the community which, I realized at a later point, provided me with the need for a reexamination of some of the academic discourses on environment. When I arrived at the meeting one early fall evening, I was surprised to find a large gathering of residents interested in finding out more about me and my interests in the region. As we discussed some of the concerns, I mentioned my own background in environmental sociology. One of the residents pointed out, "Oh, of course we care about the environment here, we always have, you know … but our concern is about our culture, our heritage, history."

I was startled by this distinction that the resident made between culture and environment, primarily because the "cultural turn" in sociology has ensured a sizable addition of literature on the impact of the cultural on the environmental. Yet, the residents' experiences with mainstream environmental organizations presented a startlingly different picture—where discourses on conservation of natural resources exclude any examination of the cultural heritages. This emerged as a central question in my research as I decided to examine some of the organizational strategies employed in constructing environments. Thus my field experiences encouraged me to reexamine some of my academic understandings of environment, development, and cultural identity.

At another level, this conversation brought home to me residents' perceptions of my position, as first and foremost, an academic "expert," someone who should be informed at the beginning of their own appreciation of the environment. In my "expert's" skin, I was the unsentimental observer, unappreciative of the community's positionality as environmentalists too. My skin was then an extendable one, representing far more than how I look or speak, but who I am in my professional capacity vis-à-vis the community residents.

My academic skin, however, was very helpful in gaining access to some of the officials working for the state agencies. My position as a researcher with an interest in starting a debate on what constitutes environment in Between the Rivers allowed me access to some of the resources of the state agencies. Officials at the Forest Service were very supportive in granting me access to documents at their libraries, encouraged me to ask questions, and generally viewed my research as an important opportunity to get their views and positions across. My "expert" status as an objective researcher thus opened doors for me to ask questions and seek explanations.

Surprisingly, even my skin color often allowed me certain advantages in my research. My location as a non-white person of foreign nationality working in a predominantly white community helped bring up issues which may not have emerged otherwise. Community in Between the Rivers has not been traditionally "mainstream" white. Instead, there are still existing small parcels of African American and Native American residents in the area. There are also stories of small segments of Chinese communities in the region in the late nineteenth century. My obvious ethnic/racial identity allowed me a peek at some of the histories of these communities and helped me raise issues of inter-racial/ethnic relationships in Between the Rivers. I was introduced to the hidden understories of the region—descendants of African American slaves still living in the area and the now-disappeared Chinese communities. During my visits to some of the existing Chinese cemeteries I was introduced to some of the dilemmas of race relations in the community. As one community resident suggested, while there are representation of various ethnic/racial groups in Between the Rivers, participation in recent struggles are not uniform. He pointed out, "We have tried to get more folks involved in the fight and have been only marginally successful … some of them (African-American residents) do show up for cemetery cleaning once in a while; but most have moved away, so getting everyone together is a problem."

While differences exist in the perceptions of "the place" in the eyes of residents along race/ethnic locations, the community came together in numerous instances to put forward their attachment to their place and history. Thus for example, former residents discussed with me ongoing coalition building initiatives with Native American residents of the area. Attempts are also being made to build bridges with the remaining African American population. One instance that I found to be of great import was the predominantly white community's role in cleaning and maintenance of the African American and Chinese cemeteries, underscoring their commitment to their land, heritage and cultural resources. Minority concerns were frequently brought up in the course of my conservations with residents generally without any prompting on my part providing me with a better understanding of the history and conflicts of the community. The conflicts and dilemmas of race/ethnic relations helped me avoid the perils of a romanticized utopian local narrative. Instead, Between the Rivers reminds one of a community just as any other, fraught with tensions, conflicts, and contradictions, yet tied together by their attachment to their land, place, and identity.

In addition to my racial/ethnic location, my gender identity allowed me a glimpse into my subjects' world (as Miriam Wells also discovered in her anthropological ethnography Strawberry Fields, 1996). My male subjects found me non-threatening in discussing "male concerns": politics of dam-building, economic advantages generated by tourism, etc., while my female respondents were more willing to share some of their understanding of changing location of women in Between the Rivers. Further, residents were also available for long discussions on the political and historical examination of their concerns. My informants were always welcoming of questions on my rather limited knowledge of American political structures and took special care to explain issues in great detail. Also, given my lack of any apparent ties with any local groups, all the groups within the region would discuss community controversies with candor. Though I did have to work on gaining trust, I believe my gender and outsider status reinforced the perception that I had no vested interest in the controversies.

Hence, during my stint in Between the Rivers, I traversed far more distances than Simmel's stranger. While I was certainly viewed as an "expert" outsider exploring some of the state-community dynamics (and it surely helped in maintaining some level of objectivity), my social locations and multiple identities complicated the picture. I was exposed to far more issues than I had hoped for in the beginning and while they helped in exploring some social and historical nuances, they also did bar me from exploring certain others. Thus, for example, the strain of displacements in Between the Rivers has contributed among other things deep divergences within the community creating structural strains in inert-personal and familial relationships. My entry into discussions of these issues was unsurprisingly limited and I had to rely on information from others to gain an understanding of these ruptures in social networks. Thus, negotiating these identities and social locations were at once empowering and alienating. Like the "marginal person" of Park and Burgess, I was constantly experiencing conflicting loyalties, changing identities, and evolving understanding of the issues at hand. Duneier (1999) refers to these shifting associations and contested negotiations aptly: "perhaps the best starting point is to be aware that a different social position can have a serious effect on one's work, and these differences must be taken seriously" (Duneier 1999: 354).

In addition, the existing sociopolitical climate can contribute to a less than conducive research environment. While it did not significantly disrupt my research, conducting research in a post-9/11 world, has had its share of limitations. My status as visitor from a "non-threatening" part of the world had to be explained at the very beginning and time and again I was neatly labeled at the beginning of a conversation. My "otherness" also led to barred doors to some community meetings, church services and similar such occasions which I accepted without rancor. Yet over all I was accepted and indulged to ask and listen as people discussed their stories and shared their histories, tentatively at first, but with increasing warmth and fellowship which I cannot but appreciate. I went there as an inquisitive, yet a nervous scholar, interested in conducting participatory research in Between the Rivers while trying to visualize and make sense of the unknown. I came away, while not with perfect "solutions," but a better understanding of not only the hardships, the strength of purpose, and the fellowship of spirit, but also of a great appreciation of the land, forests, and rivers in this remote corner of the world. On a more personal level, as my research evolved, I grew more comfortable in my skin. My shifting identities allowed me a glimpse of how ethnographer's positionality is a matter of double or some times multiple consciousness—of traversing great spaces in a short period of time both within and outside one's body and in the course of the travels allowing these multiplicities to shape both our human selves and the narratives process.

Studying the Sojourning Ethnographer: Concluding Remarks

A researcher is after all a stranger to the community, a sojourner who lives in a community for a short while engaging in seemingly "normal" activities before moving on to other commitments. She is expected to bring in a certain degree of objectivity to concerns that are emotionally charged. Yet, while she remains invested in the process, she is an outsider to the social milieu of the community, a stranger in the crowd. An understanding of this "stranger" then is important to making sense of the community that she studies. In the course of this chapter, I explored the complexities involved in conducting ethnographic field work and the dilemmas involved in negotiating the ethnographer's various commitments and identities. I explored how these conflicting emotions and contradictory locations shape and enrich both the ethnographer and the research process.

The chapter then is an exploration of the ethnographer as the sojourner in her skin. As I discussed earlier in the chapter, the ethnographer's identification as a sojourner serves two primary purposes. First, the sojourner is both a stranger to her field and an invested collaborator. While she retains certain objectivity in the research process, she is also engaged in constant negotiation with her own self-identities. Second, the ethnographer is an engaged participant. The role of the self then is an exercise in intellectual critique of the nature of knowledge (Krieger 1996). The sojourning ethnographer engages in this critique as an engaged participant in the research process. She initiates a meaningful engagement with her research site and possesses a conscious appreciation of the impact of her own locationality on the research process. The relationship between the sojourning ethnographer and her research participants is informed by an openness of identity where each enriches the other. While a temporary resident in the community, the ethnographer evolves along with the research process and her shifting perceptions shape the research narrative.

An ethnographer's travels through her skin have long been an interesting question for study. As ethnographers, we are the sojourners in our skin experiencing the shifting emotions of alienation and engagement. Like most researchers I began my study attempting to deal with the subjectivities of my own skin. Yet, the ethnographer's skin is not merely the skin of our bodies, for they extend like the boundaries of the community in which she works. As her emotional investments in the research process increases, the ethnographer travels beyond her own body, transcends her subjectivities and begins a sojourn as an engaged participant. Her skin extends to transform into the skin of the community in which she works, much like my own experience as I attempted to delve deeper into some of the sociopolitical understandings of place, environment, and cultural history. The ethnographic narrative tells us more than the story we study; it helps the narrator journey both in and out her skin and while leaving allows her to take away a tiny part of it, one that better informs her world.

NOTE

1. Articles in *Journal of Contemporary Ethnography* (August 2006) 35(4): 371–478.

Part IV

JUST PLACES: A STORY OF CULTURAL AND ENVIRONMENTAL INJUSTICE IN LAND BETWEEN THE RIVERS

Chapter 3

Cultural and Environmental History of Settlement

1779–1945

LAND BETWEEN THE RIVERS: A PLACE LOST IN TIME

Deep inside the lush green forests of western Kentucky is a place called Land Between the Rivers. The large peninsula spreads across the western parts of the states of Kentucky and Tennessee, and has long supported a self-sufficient sustainable farming community. The green rolling hills, hidden hollows, picturesque lakes, and abundant natural resources made the peninsula particularly attractive to migrants looking for rural charm and new opportunities. The peninsula's natural isolation from the surrounding region helped the community retain their frontier culture and build a close relationship with their place. Over the course of century, Land Between the Rivers slowly transformed into a place to call home. The area surrounding the peninsula was inhabited for centuries by the Shawnee and Chikasaw Indians who established extensive trade relations with the French traders in the late seventeenth century. By the early nineteenth century, the Native American tribes were forced to move westward in the aftermath of the Jackson Purchase in 1818 that granted all land west of the Tennessee River in Kentucky and Tennessee to the U.S. government (Murphree 2012). By the 1830s the area was populated by revolutionary war veterans, primarily Scotch-Irish and some German settlers who immigrated in the early nineteenth century.

As both geographic spaces and cultural sites, places are created through the use of cultural symbols that bestow and convey meaning (Greider and Garkovich 1994; Devault 1997; Davidman 1997). These cultural symbols are the essence of social interactions that result in understandings of place or create a sense of place. As Williams and Stewart (1998) suggest, a sense of place is "the collection of meanings, beliefs, symbols, values, and feelings that individuals and groups associate with a particular locality" (Williams & Stewart 1998: 19).

For the community of Land Between the Rivers, their place was more than a piece of land. Instead, it was the interpretive frame through which they defined and evaluated their own lives, and the lives of those in their communities. Sociologists have defined sense of place as not only the ability to locate things on a cognitive map, but also the attribution of meaning to a built form (Rotenberg & McDonogh 1993; Feld & Basso, 1996; Low 1992). Thus, places are social constructions that are formed when people develop a shared sense of history from collectively held stories. In this sense, the long historical settlement in Land Between the Rivers coupled with geographical isolation and frontier lifestyle created a shared sense of community and collective identity among the residents of the peninsula.

Early History of Settlement: Creating Place and Community in Land Between the Rivers

Located in the far western corners of Kentucky and Tennessee, Land Between the Rivers is a lush green inland peninsula spanning 170,000 acres of fertile land. The peninsula is enriched by two powerful rivers, Cumberland and Tennessee and spans across Lyon and Trigg counties in Kentucky and Stewart County in Tennessee. The geographical area of the peninsula is about forty miles long and occupies about 240 square miles of land (Foresta 2013, Henry 1975).

Although historical details about early settlements are rare, there is ample evidence of trade relations between the native Creek, Chikasaw, and Cherokee Indians with French and other European traders as early as 1665. European settlements began in the late eighteenth century when the Scotch-Irish Revolutionary War veterans were granted land in the peninsula as payment for their service in the war. By the mid-nineteenth century the area was fairly populated (Wallace 1992; Nickell 2007).

The relative isolation of the region allowed the frontier characteristics to linger in the area much longer than in other parts of the country. Since the primary access to the peninsula was by ferry, residents of Land Between the Rivers had minimal interaction with outsiders for six or more generations. The physical isolation contributed to a unique cultural identity among the residents who developed a unique attachment to their place and cultural heritage. The rivers were the highways in the region and ferries were integral to transportation.

The community was bound by a shared heritage, common agricultural practices, principles of environmental stewardship, and collective cultural identity. Surrounded by the rivers on three sides, the geographic isolation of the people led to the establishment of frontier settlements largely distinct from outside influences for many years. As residents settled in the land they built log cabins with windows covered with dried and stretched animal skins (Lafferty 1971, Wallace 1992). In the early years of the settlement, the staple foods of the region were "meat from wild animals, bread from corn crop, and any wild plants, berries" that were gathered through foraging. (Wallace 1992: 14). The milder climate and abundant natural resources led early pioneers to engage in bee-keeping, fermenting beverages, and the manufacture of natural byes from berries and other local fruits to sustain their families. While early settlers were largely from Europe,[1] by the turn of the century large group of migrants from Ireland, Scotland, and Northern England had settled into the region. The physical and geographical isolation of the region, smaller clan–like communities, and the widespread Celtic cultural traditions shaped the community's cultural identity as fiercely independent and proud of community traditions, heritage, and family (McWhiney 1988).

By the late eighteenth century, settlers cleared land for farming and raising livestock. Small local towns like Dover, Cadiz, Eddyville, Bumpus Mills, Hickman Creek, and Indian Mound emerged and became prosperous community towns by the early nineteenth century. While population in all three

Figure 3.1 Family homestead, Panther Creek, Stewart County (1810). Image courtesy of David Nickell.

counties in Land Between the Rivers increased substantially by the 1860s, Trigg County in KY witnessed rapid increase in population during this period. Overall, the abundant mineral deposits, fertile soil, and milder climate in the peninsula were important attractions for migration to all three counties from Virginia and North Carolina (Collins 1847; Chester 1971; Wallace1992).[2]

Surrounded by two powerful rivers, Land Between the Rivers benefited from an early history of environmental stewardship, collective attachment to their place, and shared religious practices (Parks & Folmsbee 1973, Hamer 1933). Most of the settlers worshipped in the local Baptist churches established in the peninsula, though a local Catholic Church was later established in 1900 for newly arrived German immigrants seeking a place to worship.[3] The fertile land and abundant natural resources offered the early settlers ample land for sustainable farming, livestock grazing, and supplemental hunting to meet community needs. By the early nineteenth century, large landowners in the peninsula were planting corn, hemp, cotton for sale, and for supporting their own families (Wallace 1992: 34). The labor-intensive nature of cotton production led to a gradual shift toward tobacco cultivation. The close proximity of the two rivers to the peninsula was advantageous in transporting crops to other parts of the country. For most of the local farmers though, the economic activities in their farms were solely directed toward meeting their own family needs. In fact, for many years the community's economic needs were solely supported by their collective farming and grazing initiatives practiced in the large tracts of rugged land locally known as the *Coalins*.

The Coalins refers to "commons" land left unused due to their rugged and hilly terrain. None in the community had any legal claim to the land and the 37000 acres of the Coalins area was open for collective use. As a former resident-scholar pointed out,

> The Coalins was so rugged, with deep folds in the hills and both rock and ore protruding from the thin soil, that the original settlers avoided it. The result was that as communities built up, they ringed this rugged terrain with no one having filed a claim to it. (Nickell 2007: 7)

Table 3.1 Population of Three Counties in Land between the Rivers (1850–2000) (U.S. Census Bureau)

Year	*Lyon County Population*	*Trigg County Population*	*Stewart County Population*
1860	5,500	11,000	9,000
1900	9,000	14,000	15,000
1940	9,000	12,000	13,000
1960	6,000	9,000	8,000
2000	8,080	12,370	12,597

The rugged land remained barren for decades and was primarily used for hunting and pasture by the community who managed it carefully for their future (Henry n.d.; Nickell 2007; Wallace 1992, Montell 1975). The management and protection of the Coalins showcase early community history of environmental sustainability. Under the management of the residents of Land Between the Rivers, the Coalins served as home to many species of animals that became extinct from excessive hunting. Thus for example, the community was able to protect wild turkeys in the Coalins at a time when they were extinct in Kentucky.

Along with their environmental stewardship practices in the Coalins, the community's sense of place emerged from their cultural and religious traditions. The churches and the cemeteries that dotted the peninsula formed the early historical markers in Land Between the Rivers and formed the essence of the community's cultural heritage. Pioneer settlers in the area were deeply religious and brought their unique religious traditions to Land Between the Rivers. Most of the local churches, including Dry Creek and Crockett Creek churches were Baptists and were established between 1798 and 1810. Two Catholic churches were also built in the late nineteenth centuries. Other denominations were also represented in the region. The community organized religious debates between various denominations that promoted community solidarity and were great sources of entertainment (Henry n.d.).

Along with the churches, the peninsula was dotted with over two hundred family cemeteries, many of which still exist and are managed by the former residents, who regularly volunteer for spring cleaning initiatives. Additionally, the cemetery sites are often used for fetes and book fairs. Describing the importance of the cemeteries in the reconstruction of a coherent community identity, one community member pointed out:

> The cemeteries are central to our life as a community of Between the Rivers. Those long gone people and their history remain a thread that binds us all together … the act of working in the cemeteries [weekly cleaning during the winter months] with other former residents, many of us only recognize each other by family names, helps to re-establish connections among individuals and families. (Female, former resident, homemaker, age 50)

The cultural identity of the community of Land Between the Rivers was informed by visible physical markers as well as stories and memories of the long history of settlement. The community heritage was shaped by the early settlers' independent frontier lifestyles, environmental stewardships, sustainable farming, and religious practices. The isolated geographical landscape fostered a sense of insular community identity that survived much of the economic and political turmoil of the nineteenth century and the early decades of the twentieth century. Change however came to the region in the period following the Great Depression era when socioeconomic development plans were implemented under the New Deal. The New Deal began a process of socioeconomic development in the U.S. South that had a major impact on Land Between the Rivers. As a resident observed:

> My grandfather would recite poetry to us, often his own. He would tell us the family stories. He would expound on the history of the farm and how we should improve it for our children and grandchildren. We heard it all many times. We were never told, and did not need to be told, that the farm was our "place." Our place was an integral part of [Land] Between the Rivers that has remained only superficially touched by the outside civilization. We saw it as home—where outsiders might visit, but where we belonged. It was also unspoken, but clearly understood, that the commonly heard phrase, "Between the Rivers" was inclusive, encompassing not merely our family and farm but the neighbors and their places. (Nickell, David L [n.d.] "A Story of TVA's Removal of Families from Between The Rivers." http://www.betweentherivers.org/dnickell.html)

Industrialization in the Mid-Nineteenth Century: Impact on the Residents

The economic development of the region in the mid-nineteenth century brought remarkable changes to the management and protection of the Coalins. As demand for iron ore grew nationwide, investors from the North flocked to the region and filed legal claim to the Coalins for mining iron ore (Hogan 1971, TVA 1983). The Watson and Hillman furnaces that were built in the mid-nineteenth century emerged as important players in shaping the economic destiny of the region. The relative ease of access provided by the two rivers quickly transformed Land Between the Rivers and its surrounding communities into the nation's leading producer of iron ore. Historical records narrate these transformations: "Stewart County alone, in the southern portion of LBL, had 14 operating iron furnaces. These were massive stone and wooden structures which, fed by charcoal made from local timber, produced enough "blast" to fuel the smelting process. By 1830, the region was the third-largest producer of iron ore in the country" (U.S. Forest Service [n.d.]).

The rising demands of iron ore during the Civil War catapulted the region from a predominantly agricultural settlement to a commercial hub supporting the railroad industry in the mid-nineteenth century. In fact brown iron ore was abundant in the land and early settlers documented it as early as 1796. However, it was only by 1820 that regional interest in the mining of iron ore increased.[4] Together with the abundant wood fuel, accessible waterways, and availability of limestone for processing the ore, Land Between the Rivers offered early entrepreneurs important opportunities. At its peak, there were fourteen iron furnaces located in Stewart County, TN (now part of southern

Figure 3.2 Old furnaces in Land Between the Rivers. Image courtesy of David Nickell.

LBL).[5] Two of these furnaces remain intact today (Great Western Furnace and Center Furnace) and are located along the Trace[6] in Land Between the Lakes.[7]

Cumberland Iron Furnace, one of the largest in Stewart County, began production in the early 1820s. The company owned over 60,000 acres of land on both sides of the Cumberland River and worked in close partnership with other local partners including owners of the Bear Spring furnace, Dover Furnace, Center Furnace, Saline Furnace, and others (Wallace 1992: 38). The workers employed by the iron ore industry were diverse. In addition to the white furnace workers, immigrant Irish workers, and African American slaves, Land Between the Rivers also became home to a large group of Chinese migrants employed as "Coolies" in the region.

While an accurate accounting of the exact number of furnace workers is not available, by a general estimate several hundred Irishmen were living and working in the furnaces by the mid-nineteenth century. A large number of African American men workers in the furnaces as well and most were slaves owned by iron masters or farmers in the area. In the post-Civil War era, Chinese laborers were brought to work in the furnaces, though some of these laborers came to the peninsula as early as 1820. Most of these Chinese workers were segregated from the white furnace laborers and received lower pay. Many stayed for brief periods and most returned home or travelled further west to work on the railroad industry. Some of the Chinese workers stayed back and brought their families to the region. While scanty records are available for these newer migrants, a sole Chinese cemetery still stands on the edges of Land Between the Rivers as a silent marker of the rapidly transforming economic and cultural landscapes of the United States. As demands for iron ore grew, the community lost much of its access to the Coalins and environmental stewardship initiatives were severely curtailed.

The meteoric rise of the demand for iron ore was relatively short-lived. Fear of slave insurrections and the looming civil war reduced production in the region. As the war ended demand for iron ore never picked up, and businesses shifted their base further South. The economic changes left the community relatively unchanged. The residents were largely engaged in farming with little participation in iron ore mining in the region. In fact, for much of the nineteenth and twentieth centuries, the farmers were largely unaffected by the ongoing industrial developments in the region. Agriculture, livestock, and subsistence farming were always the most important occupations in the area and employed most of the residents living in the region. In particular, the settlers were engaged in the production and marketing of tobacco, beef, and pork, whiskey, and timber which were produced in abundance and transported nationwide via the rivers (Nickell 2007). As the furnaces were closed at the turn of the twentieth century, the residents again stepped in to return the Coalins to its former state by encouraging livestock management and protection of wildlife in the Coalins.

Returning to Nature: Community Environmental Stewardships in the Early Twentieth Century

The success of the community's environmental ventures led to the creation of the Hillman Game Refuge in the Coalins in 1919 by the state of Kentucky. The overarching purpose of the refuge was to facilitate and promote ongoing wildlife conservation projects that were started by the community residents (Henry n.d.; Nickell 2007). As part of this initiative, the residents set up a semi-formal regulatory system of livestock management in the Coalins that was patrolled by local farmers trained as game wardens. Further formalization occurred in 1935 when the Department of Interior (now known as the U.S. Department of Fish & Wildlife) took control of the refuge and renamed it Kentucky Woodlands Wildlife Refuge. The refuge continued the community-initiated successful management of wild turkeys and deer, whose population had previously dwindled in the state.

With the transfer of land under eminent domain in 1965, TVA declared its decision to claim the Refuge which was rechristened as the Environmental Education Area managed by the Land Between the Lakes. In recent years, the Forest Service has been lobbying to convert this to an Oak-Grass Habitat and a Wildlife Viewing Area. The formalization of the environmental stewardship program in the Coalins was accompanied by the non-recognition of the early environmental history in the peninsula and the role of the community in the protection of the wildlife in the region. No formal mention of the Coalins can be found in any official maps of the Land Between the Rivers and exists only in the collective memory of the community.

The early years of the settlement, community life, and the subsequent displacement shaped the community's collective identity in important ways. This chapter narrates the community's early history in Land Between the Rivers and explores how a sense of place attachment and cultural identity emerged over the course of two hundred years. More specifically, I examined how early histories of settlements were enriched by community's environmental stewardship in the region and cultural and religious activities that shaped their collective identity. I also examine the impact of socioeconomic changes in the region in the mid-nineteenth century on the community as iron ore mining catapulted the region into the national map. Finally, I examine how TVA's economic development projects radically transformed the community in the aftermath of their displacement. I explore the emergence of a nascent environmental and cultural justice movement in Land Between the Rivers as the community resisted the challenges to their collective identity, history, and cultural heritage.

The next chapter, *Development Projects and Displacement Narratives: 1945-present,* discusses the genesis and development of the ensuing phase in the environmental justice movement in Land Between the Rivers. The anti-commercialization protests initiated a movement for economic justice for the residents which was followed by calls for institutional and political justice in recent years as the community flights for greater inclusion and participation in cultural and environmental decision-making in the region. The post-1965 period in Land Between the Rivers is marked by a demand for greater agency for the former residents through legal and political channels as the community attempts to seek legitimation for their struggle for cultural and environmental rights.

NOTES

1. Many of the early settlers were migrants from Germany, Switzerland, and Netherlands.
2. http://www.betweentherivers.org/pinegarletter.html
3. https://www.explorekentuckylake.com/explorations/Ststevens.htm
4. Stewart County Court Records.
5. Many of these furnaces were operated by regional (non-local) companies who shipped the iron ore to the development of the railroad industry.
6. The Trace was originally known as State Route 49 that runs through much of the Tennessee portion of Land Between the Lakes.
7. Great Western Furnace is located near the former site of Model, TN, a township in Land Between the Rivers. In fact, the furnace is all that remains of the town of Model, TN. The LBL official map (see below), locates the ruins of the Great Western Furnace near *The Homeplace*, close to the Kentucky/Tennessee border.

Chapter 4

Development Projects and Displacement Narratives

1945–present

ENVISIONING SPACES AND RECONSTRUCTING PLACES

Places are shaped by centuries of memories and histories that give meaning and a sense of collective belongingness to a community. Sociologically speaking, a sense of place is not only the ability to locate things on a cognitive map, but also the attribution of meaning to a built-form or natural spot (Rotenberg & McDonogh 1993, Walter 1988). Places are made as people ascribe qualities to geographical spaces. They are also constructed as people develop a shared sense of history from collectively shared memories about their communities.

As both geographic spaces and cultural sites, places are important cultural symbols that bestow and convey meaning (Greider and Garkovich 1994). These approaches to place and cultural identities have been explored in sociological studies on cultural politics of place and displacement. Researchers have examined how biophysical nature is socially constructed (Gupta and Ferguson 1992), studied place attachments and its effect on collective identity (Tuan 1993, Williams and Carr 1993, Altman and Low 1992), explored connections between a communal place and environmental values (Norton and Hannon 1997), and analyzed cultural landscapes and other geographical spaces (Feld and Basso 1996). Yet, place research assumes significance in narratives of displacement where communities' sense of collective well-being is severely tested.

In Land Between the Rivers, this sense of security and well-being associated with a community *in place* suffered severe strain as commercial development began in the region with the adoption of the New Deal and the introduction of a federal mandate for national reconstruction. The history of the displacement in the Land Between the Rivers dates back to the

New Deal that stated the urgent need for economic reconstruction in the U.S. South, severely impacted by economic devastations of the Great Depression. As part of the New Deal, the federal government established the Tennessee Valley Authority (TVA) to revitalize the national economy and established a series of economic developmental projects. Land Between the Rivers was identified by the TVA as the site for one of the many hydroelectric power projects to generate jobs and produce electricity to meet the needs of a growing economy. The ongoing World War II slowed the construction process significantly.

Construction of the Kentucky Dam began in 1938 on the Tennessee River, followed by the Barkley Dam on the Cumberland River in 1960. As work progressed, residents were moved to higher ground in five phases spread over a period of over two decades. Following the completion of the Barkley Dam in 1966, the Kennedy administration ordered the remaining land to be converted into a federal refuge for promoting environmental conservation in the region. Following this decision TVA decided to remove the remaining families from Land Between the Rivers, and the confiscated land was turned into a nature preserve, officially known as the Land Between the Lakes.

Figure 4.1 Construction of the Barkley Dam. Image courtesy of David Nickell.

Dam construction in Land Between the Rivers began in earnest in the late 1930s as TVA unveiled its plan to build two dams on the two rivers—Cumberland and Tennessee—that encircled the community landscape. Construction of the Kentucky Dam on the Tennessee River started in 1939. In the mid-1950s United State Army Corps of Engineers began surveying land around Eddyville, KY for a second dam project. Construction of Barkley Dam on the Cumberland River, located two miles from Kentucky Dam, began in 1959 and was completed five years later.[1] In the intervening years several local communities including Kuttawa, Eddyville, Cadiz, Model, and Gatlin Point were displaced in successive phases as residents were moved away from flooded riverbeds disrupting community homes, farms, and lives. In fact, as residents attest and archival reports observe, these dam construction projects permanently flooded the best cropland on the peninsula and involved three rounds of displacements using eminent domain, disrupting social organizations and farming activities in Land Between the Rivers. Finally, after the completion of the dams in 1966, TVA decided to remove the remaining families from Land Between the Rivers and using eminent domain turned the remaining land into a nature preserve, known as the Land Between the Lakes.

A Community in Crisis: Geographical and Cultural Implications of Displacement

The construction of the dams and the federal refuge had significant impact on the community's claims to place and cultural identity. It not only led to the physical displacement of these tight-knit families and neighbors, but also led to a crisis of identity and sense of place and resulted in the loss of heritage, history, and identity, and the forced displacement serves as a reminder of injustice. The history of the forced displacement in Land Between the Rivers is a long and tortuous one and was consistently resisted by large segments of the community. It began in 1938 when TVA started the construction of the Kentucky dam in the Cumberland River. Using eminent domain, TVA removed a number of families from the region and bought their most productive farm land. The phased removal continued over the next two decades as TVA built a second dam followed by a natural preserve in 1965. As residents were forced off their land, TVA destroyed churches, homesteads, farms, and all relevant cultural artifacts of the community.

Community residents were fiercely opposed to displacement efforts that would severely affect their living conditions and long historical association.

Mobilization efforts in Land Between the Rivers began soon after the construction projects were unveiled. While the onset of World War II severely affected community efforts, protests intensified soon after. Residents organized sits-ins, marches, garnered signatures, and demanded a congressional hearing.

The community protests were frequently labeled as opinions "on the fringes" and residents were stereotyped often as "backward people" and "standing in the way of progress." Referring to TVA's stereotyping of the community, a congressional report states,

> TVA has used every available means of destroying the morale of the people [of Land Between the Rivers]. Months before this project was officially in operation, TVA directors and other officials spoke to groups in many surrounding towns, but refused to talk to the people involved. ... By sly innuendo, ridicule, false accusation, and outright vilification our people have been pictured to the entire nation as nothing more than illiterate outcasts. Now, a marvelous transformation has taken place. By the simple transfer of title from the native landowners to TVA, this "ugly, garish, impoverished strip of land" has become a "beautiful land of enchantment." (Stubblefield Report 1968)[2]

Marginalization and belittling of community concerns were common among officials in both the TVA as well as the U.S. Forest Service. During one of my conversations with the officials in Land Between the Lakes, one participant observed:

> I think the challenge is that it isn't just the former residents, I mean I have worked long enough at the forest service to be comfortable enough to say this, you know, the public has a very diverse view of management of public land, and when you have that much diversity, trying to satisfy all is not possible. And so, there will always be those on the fringes that will disagree with us and so I think you add another layer to that with the former residents—so we tend to hear more from the ones that disagree with us than the ones that do agree with us.

This was also evident in the attempts to ignore the long-established sustainable relationship of the community with its land. In conversations during my research, residents resisted the attempts to paint the community as poor and desperate and in grave need of rescue from their despair. Instead, community residents shared anecdotes about the comparative ease with which residents survived the depression years, largely due to their ability to support themselves by working their lands. Residents like David Nickell shared stories about established environmental practices like the collective regulation of commons property used for grazing. Locally known as *Coalins*, the commons land was long held in Kentucky as an outstanding effort to protect native species including wild Turkey ([Redacted]). The popularity and success of the commons endeavor led to the creation a more formalized wildlife and game management program at the turn of the twentieth century.

Yet, TVA's narrative offers little acknowledgment of the centuries-old efforts to promote environmental stewardship in Land Between the Rivers.

Instead, through repeated stereotyping of residents as poor, backward, and lacking rudimentary understanding of the economic opportunities from dam construction, TVA successfully eradicated any claims to environmental and cultural ownership to residents. One former resident reminisced about the abundant use of negative stereotypes of the community in TVA's propaganda movies featuring the depressing and deplorable conditions of the community:

> When I was ten, relatives from out of state came to visit. I was informed that we were "going over home" to let them see what the place looked like now. For the first time we went to the visitors information center at LBL. There was a prominent display there representing the "twixt the river culture." The display consisted solely of a moonshine still and a picture of someone sitting in a corn-crib husking corn. The room full of tourists gasped at the desperate condition of the backward people living in the area. The house they showed as "typical" of the dwellings of these pitiful people was, according to my mother, somebody's henhouse. The film told of how grateful the people were to be released from their Appalachian-like poverty and failed culture. I left the building in tears. (Nickell 2001: 8)

While various incidents during the displacement showed TVA's efforts to stereotype and marginalize the community claims for justice, residents felt the injustice of displacement most clearly when TVA decided to demolish the churches and cemeteries. Interviews with residents point to the unjust and cavalier attitude of TVA during the displacement that, "clearly established an ugly pattern of unethical and unjust acts used so extensively as to appear as unquestionable policies of the agency."

The local churches have always existed as a prominent collective space that epitomized the collective sense of place and history among residents. Churches have extensively served the community since the early decades of local settlements. Spread across communities of Golden Pond, KY Model, TN, Tharpe, TN, and Bethlehem, KY, among others, these churches offered a communal place to residents to worship as well as share their collective cultural heritage.

As dam constructions began, TVA asked the residents to move the churches off the land at their own expenses along with the rest of their property. While many of the churches were submerged in the rising water, others were bulldozed by the TVA along with abandoned homesteads and other local artifacts. Recalling the negotiations with the TVA, one former resident observed:

> We went to the TVA with our request [to protect the churches]. They said they wanted the churches moved off the land at our own expense. Our choice was clear, you know, move the church, or they will bulldoze them. Most of us had

> to move themselves, their houses, property with them. … Moving the churches were never an option—most were destroyed.

Adjacent to these community churches were the two hundred family cemeteries which the TVA also wanted to remove from the refuge.

Community resistance forced them to abandon the plan and the cemeteries remain a grim witness to the sad past. The residents remain deeply committed to continued renovations as well as maintenance of these artifacts. Despite the removal of the residents, the community residents were collectively granted ownership and burial rights in the cemeteries by the federal government (Nickell 2007). In fact, these community stewardship initiatives have sustained through the TVA years and continued after the management of LBL was transferred to the United States Forest Service on October 1, 1999, under the Land Between the Lakes Protection Act of 1998. Describing the importance of the cemeteries in the reconstruction of a coherent community identity, one resident noted:

Figure 4.2 Tombstone in Cemetery in Land Between the Lakes Refuge. Image courtesy of David Nickell.

> The cemeteries are central to our life as a community of Land Between the Rivers. Those long gone people and their history remain a thread that binds us all together … the act of working in the cemeteries [weekly cleaning during the winter months] with other former residents, many of us only recognize each other only by family names helps to re-establish connections among individuals and families.

However, under the new management, controversies over access rights to the cemeteries emerged again in 2001. The "Cemetery Handbook" issued by the Forest Service presented a set of legal directive to the residents, regulating access rights of locals to the cemeteries on federal conserved land, arguing that only groups with "legal" access to cultural sites under the Native American Graves Protection and Repatriation Act (NAGPRA) are exempted. The proposed handbook acknowledged the community's access to the cemeteries with the caveat that the guidelines will allow the agency to regulate those rights (Banerjee & Steinberg 2015).

Residents mobilized to challenge the infringement to community's cultural rights and were able to successfully negotiate the withdrawal of the handbook as well as the continued enforcement of the original contract about community ownership of the cemeteries. The conflict over the cemeteries reflected the broader efforts to manage and regulate the community's rights to their own cultural identity. It fostered growing perceptions of injustice among residents as they continued to organize and mobilize their community resources to seek key opportunities to share inputs about cultural resources in the refuge and influence federal decision-making in the refuge. Community resistance to these regulations also allowed residents to reaffirm their connections to their history, seek more authentic representation of their displacement narratives, and continue to forge ties and networks with environmental and cultural activists in the region.

The history of mobilization in Land Between the Rivers assumed prominence even before the conflict over cultural rights to the cemeteries. In the late 1950s, as a second dam was in planning stages on the Cumberland River, TVA, with support from the Kennedy administration, developed a plan to turn 170,000 acres of land in to a federal refuge as promote it as a "green magnate" to drive economic development in the region. Over the next few decades, shrinking budget was coupled with a lack of a vision plan among TVA officials responsible for promoting better management of the refuge. As a result, LBL under the TVA seemed to be "stranded, ill equipped in the world around it, when the world of progressive optimism ran out" (Foresta 2013: 187). In the 1990s, TVA decided to develop a new plan for LBL's future. In December 1995, TVA proposed an economic development plan called the *Five Concepts* aimed to replace lack of growing employment as originally promised as well as growing budget constraints and shrinking federal funding. The most ambitious (and equally outrageous) elements of the plan included two resorts, a conference center, two campground resorts, a starlight theater, beach parks, marinas, and a heritage theme park.

The economic development plan in LBL was, however, in direct violation of its original promise to the community. In the late 1950s, when the federal government decided to build the refuge and entrusted the TVA to oversee its development, the agency had promised to ensure continued non-commercialization of the land. This promise remained enshrined in the LBL Protection Act and referenced in community narratives:

> The intent of the LBL Protection Act was to insure LBL's future as a National Recreation Area. The Protection Act was intended to provide adequate federal funding for LBL and to codify in law the original mission and promise made to approximately 949 families who through eminent domain forfeited their homes, churches, businesses, and way of life in order that a National Recreation Area void of commercial development might be established. (USFS 2004: archival source)

As budget constraints decimated available federal funding and earnings from tourism did not sufficiently materialize, TVA began to develop a new plan to generate funds from economic development in the refuge. In 1995, it proposed a recreational project in the region that included plans for waterfront villas, condos, and marinas, construction of a theme park, and large-scale increase in timber sale from the forests. For the community, TVA economic development project was widely viewed as a violation of its original promise for non-commercialization of the land. The forced displacement of the community over the decades with minimal compensation was only made somewhat meaningful by the solemn promise by the government to protect the land from commercialization. Hence, the *Five Concepts* plan was viewed as unlawful and a clear violation of the negotiated agreement between the TVA and the community. This feeling of betrayal bolstered the mobilization efforts among former residents who received widespread support from local communities, hunting groups, regional environmental groups, and even the Audubon Society. The community residents demanded that,

> TVA honor their word. They told us that we have no standing with regards to the situation. The resource manager told me I had no more claim to our family cemetery than any other citizen. We were told that the promise made in exchange for the taking of our homes, farms and heritage was no longer valid. Due to the intense outcry over the proposals, a congressional hearing was held on the fate of LBL. I testified at that hearing that we still have a cultural connection to our place that we intend to maintain and that we should have standing with regard

> to what happens to places and things significant to our cultural identity. I asked the congressional members in attendance, "How many generations do a people have to live in one place before they are native to that place?" No answer was ever given. (Nickell 2001:9)

To resist these plans, the former residents along with interested groups in the region formed a grassroots coalition called the Concept Zero Task Force. The purpose of the group was to forge a platform to share community concerns with the larger public (Daniel 1997; Nickell 2007). The public support that the coalition received led to the shelving of the contentious economic development plan and led to a congressional decision to transfer the management rights for LBL to U.S. Forest Service.[3] The Forest Service suspended large-scale economic development, though community members have consistently protested ongoing timber harvesting in the refuge, lack of community engagement on environmental projects in the area like the elk-bison prairie (Banerjee 2012). As a community resident states,

> You need to understand ... while we agree that some logging may be needed, that TVA identified specifically was the place where we have done conservation work for a very long time. I guess, you know, it seemed like they we trying to do away with all evidence of our environmental contributions, while making money off it.

Continued concerns since the transfer to the Forest Service suggest that the community's struggle for justice has evolved over the years. The question that shapes mobilization efforts in recent years emphasizes the need to seek a "place at the table" by gaining legal recognition for the community's cultural history and identity. As one resident reminisced,

> We are still resisting the Forest Service's attempts to keep the commercial developments TVA put in place prior to the transfer. More is at stake now than just whether the promise will be kept—even though that promise has always been the lynch pin of our shared experience. Our claim to a unique relationship to our place is now up for grabs. We have no official recognition and thus no legal protections for our cultural properties. Our struggle is only beginning.

Thus, Land Between the Rivers witnessed a symbolic transformation in the hands of TVA. While this process was a result of political decision to displace people in the name of economic development for the "greater good" of the nation-state, it also draws our attention to the lack of attention to the impact of displacement on local communities and their cultures (Roy 1999). In the process of constructing economies of scale, acts of environmental and cultural injustice in Land Between the Rivers became an effective tool for state legitimation at the cost of displacements of histories, stories, and memories.

With the entry of U.S. Forest Service in 1999, a new era of land and forest management practices have emerged. While policies under TVA were guided by top-down modernist traditionalism, the Forest Service has set up a more laterally managed administrative unit whose policies are couched in more benign and inclusive terms. These are particularly evident in three contentions concerns since 1999: *cemetery regulation and maintenance, preservation of a surviving church*, and finally in *heritage management* plans. While in each case Forest Service was more amenable to more inputs from all interested groups, significant concerns still remain with community engagement, and actual adoption of inputs from formal residents. As the cemetery management conflict was discussed earlier, I discuss the claims of injustice in community concerns about the two remaining issues—the ownership of a particular cultural artifact and the planning of a heritage management resource center.

The fate of the numerous churches in Land Between the Rivers has always been a contentious issue in the region. Most the churches were bulldozed and destroyed during the displacement years as TVA was unwilling to offer resources for relocation. However, a sole Catholic Church, St. Stephen's Church, remained in the forests largely due to its remote location and has been preserved and maintained by the community for many years. Early German immigrants to the area built the church in 1900 and regular prayer services were held until 1945.[4] For the next two decades, the church sat vacant next to the church's cemetery and was largely ignored during TVA acquisition of the land for the refuge. This was partly due to the church's remote location, about 2 miles from the nearest highway and due to flooding of much of the nearby land due to dam construction. St. Stephens's church remains one of the earliest and still existing cultural artifacts from the settlement years. Working with the U.S. Forest Service, the former residents successfully lobbied the State Historic Preservation Office (SHPO) to list the church as a heritage site. The community was also able to finally include a link to the church in the Forest Service maps and information desks though disagreement remains about the preservation and maintenance of this and other artifacts.

The restoration of the St. Stephens's Church led to growing introspection among community members about the need for a broader community engagement in the preservation of oral histories, artifacts, and other cultural heritages in the refuge. These efforts assumed great importance in recent years over a Heritage Management Project in LBL.[5] The project aimed to protect and promote local cultural heritages and artifacts available in the refuge through predictive modeling and site identification tools. The project proposal also presented several other tools to promote environmental and cultural values including, allocating more resources for cultural and environmental education, effective monitoring of local sites, establishing a set of

Figure 4.3 Old St. Stephens Catholic Church. Image courtesy of David Nickell.

protocols for discovery of historical sites and remains, and creating an effective theft prevention system (Banerjee 2014).

While the community shares Forest Service vision of heritage they differ on two significant issues: a) on how the term is used to determine “heritage” in Land Between the Rivers and b) who gets to participate in the decision-making process. Interviews with former residents suggest great concern and mistrust regarding both issues. To begin with, there remains a deep disconnect in the perception of “heritage” in the eyes of the state agencies and former residents—while the former explore “heritage” in an *archaeological* sense, as documentation of coins, silverware, plates, bridles, and horseshoes, the latter define heritage in a more *anthropological* sense, as cultural histories, stories, memories, and identities. Neither can be dissociated from the other and if viewed together may help shape the cultural uniqueness of community life in the refuge.

Second, issues of participation remain an important concern as well and have informed how the community and management perceptions diverge on the specific nature of participatory decision-making. For example, in 2003, the U.S. Forest Service sent out a call for participation in the heritage identification project.[6] As the notice pointed out,

> It would be good to know who would like to contribute specific heritage site information during the individual open land field surveys. For example, you may not have any interest in what kind of plan I draw up to do the survey but do want to make sure that when I am out surveying fields that you or your family used to live on that you can share information with me or know what I am finding on the ground.

The notice further calls for more information on stories, and heritages in the region,

> And lastly, as usual, I am also requesting help in identifying any historic or prehistoric sites, artifacts, or features as well as stories that may be within or associated with the proposed project areas from those of you who have expressed an interest and/or concern for heritage resources at LBL. This information will help and be incorporated as the surveys progress.

While this can be viewed as an important step in acknowledgment the cultural identity and place-based heritage in Land Between the Rivers, it still leaves open the question of the nature and scope of participation in the project. As one community resident pointed out,

> There are a couple of important factors reflective of the ongoing lack of agreement. We are *not* actually consultants. We are listed, individually, as “interested

> parties." A tourist from Michigan can be an interested party, which means you get notified of relevant projects and events and you can submit information you think may be relevant to what they are doing. Being a "consultant" is a formal designation that indicates it is recognized as *your* heritage they are impacting and you are being consulted to see 1) what is important from *your* perspective, and 2) how you would like to have the situation handled. The tribes get that designation automatically; we don't.

Further, the call for heritage protection lies in what the community considers a misrepresentation of the history of Land Between the Rivers. As a resident succinctly points out, "If you notice in that scoping letter it makes a point of their intention to protect and promote 'LBL's rich heritage.' LBL doesn't have a rich heritage! It is only 40 years old. *We* have a rich heritage, which they are attempting to usurp by simply leaving us out of the official documents and processes."

Notwithstanding these conflicts, the community residents have achieved significant success in staking their claims for cultural and environmental justice. They sought and were granted signatory party status in the heritage management project in recent years and have been effectively engaged in collaborative actions with other interested groups.

The struggle for environmental and cultural justice for the community of Land Between the Rivers has come a very long way since the displacement of the residents in the mid-twentieth century. While the journey has borne mixed results from a movement perspective, it offers critical tools to understand how environmental justice movements themselves are often complex narratives that intersect in surprising ways with various expressions of identity. Thus, while economic injustices have long occupied the center stage in environmental justice debates, other less obvious forms of injustices, including cultural identity, place politics, and institutional fairness remain critically important. As I discussed in the earlier chapters, environmental justice is best conceived as a repertoire of critical justice lenses that needs greater theoretical engagement. In particular, I identified four strategies or tools of justice—cultural justice, economic justice, processual justice, and political justice—as part of a conceptual analysis of environmental justice.

In the next chapter, The Environmental Justice Movement in Land Between the Rivers, I explore how the environmental justice movement emerged in Land Between the Rivers and developed over the course of the years. Residents adopted plural protest strategies to seek justice and each of these justice strategies were employed in mobilizing resources for the environmental justice movement in Land Between the Rivers. Early efforts to mobilize residents largely focused on calls for cultural justice as residents strived to protect their cultural history and heritage in the region in the face of dam and refuge construction initiatives by the TVA. In the 1990s as the focus on development shifted and financial demands on the TVA led to economic redevelopment plans, protest strategies were focused on economic justice. Further, in the early 2000s the community demanded greater access to institutional resources to demand institutional justice and a place at the negotiation table for continued preservation of environmental and cultural resources. Finally, residents have been involved in demanding political justice through the construction and framing of an alternative narrative that is owned and shared by the residents.

NOTES

1. https://www.explorekentuckylake.com/lakesarea/history/barkley.htm
2. Congressman Stubblefield's report to the Congress referred to some of the high-handed tactics employed by the TVA in Between the Rivers. Originally a supporter of the TVA project, residents' concerns led him to collect comments from the community and formulate a congressional report to address some of these allegations of injustice. As a result of the report, some of the displaced communities were able to receive compensations from TVA.
3. The Concept Zero Task Force, along with former residents preferred the National Park Service as the new management.
4. https://www.explorekentuckylake.com/explorations/Ststevens.htm
5. http://www.achp.gov/docs/tn-ky.fs.land%20between%20the%20lakes.pa.apr14.pdf
6. https://www.landbetweenthelakes.us/wp-content/uploads/2014/05/2003HeritageResMgntPlan.pdf

Part V

CONTEXTUALIZING THE ENVIRONMENTAL JUSTICE MOVEMENT IN LAND BETWEEN THE RIVERS

Chapter 5

Exploring Strategies of Justice in Community Mobilization

The Case of Land Between the Rivers

FRAMING THE MOVEMENT IN LAND BETWEEN THE RIVERS

The environmental justice movement in Land Between the Rivers presents a unique opportunity to explore how diverse sets of movement tools adopted over the course of a movement shape collective mobilization for environmental justice. The community of Land Between the Rivers in rural western Kentucky tells us the story of a displaced rural community's struggle for cultural and environmental justice over several decades. The community was displaced in the post-New Deal era when economic revitalization programs under the TVA led to construction of dams across the Southeast. Though the displacement by the Tennessee Valley Authority (TVA) occurred in the mid-twentieth century, the community has been fighting for non-commercialization of the land, recognition of their cultural rights, access to institutional decision-making, and protection of their history and heritage. In the course of their movement, the community has invoked multiple movement strategies to justify their claims, mobilize resources, and seek a prominent platform for greater acknowledgment of their rights to a cultural heritage. The movement also embodies the need for a more complex and multidimensional approach to what essentially represents the "justice" in environmental justice.

Environmental justice movements have long incorporated a diverse set of protest tools to mobilize communities. These tools are not limited to the call for fair economic distribution of resources as noted by scholarly studies on environmental justice movements. Instead, environmental justice protests include narratives on cultural heritage, strategies for promoting greater institutional access to affected communities, and a platform to promote alternative visions for preserving collective identity and the pursuit of justice. These developments are increasingly incorporated in studies on environmental justice. For example, cultural geographers have examined the spatial politics of urban environmentalism, gendered nature of justice questions, and discourses on the racialized state in shaping environmental justice conflicts (Walker 2009; Buckingham et al. 2009; Kurtz 2009). Other scholars have explored alternative traditions of Actor Network theory, and the discursive politics of environmentalism to redefine environmental justice (Holifield 2009; Schlosberg 2007). Researchers have also examined how current discourses on climate change need to be informed by distributive and institutional barriers to a cooler planet (Checker, 2010, Dorsey, 2007, Roberts & Park 2009, Stallworthy 2009).

Despite these important developments, environmental justice studies stand on the threshold of new global sociopolitical and environmental realities. The political economic shifts toward a global marketplace have unleashed a global trajectory of toxic transfers resulting in significant resistance from displaced communities across the world that frequently links environmental devastations to violations of human rights. These emerging conflicts over the egregious violations of human rights highlight the need to conceptualize environmental devastations in the parlance of human rights violations. This chapter argues that expanding the conceptual parameters of environmental justice creates remarkable possibilities for theorization and allows us to address growing violations of environmental rights around the world.

Second, environmental justice conflicts occur in place and the loss of collective cultural identity and heritage frequently inform these struggles. Cultural contexts are neither static nor measurable in economic terms. Instead, they employ symbols and imageries that capture histories, memories, and attachment to places. These claims form the understories in conflicts over loss of place in the Amazon, the loss of livelihood among the nomadic snake charmers in rural hinterlands of India, and finds vivid expressions in Native American conflicts for place in western United States. Since cultural imageries and symbols are not themselves things, they are not distributable things. As such, they cannot themselves be captured adequately by the predominantly distributive approach to environmental justice. Any definition of environmental justice thus needs to incorporate these cultural understandings of place, memories, histories, and the loss of cultural identity.

As environmental justice movements spread globally, researchers have noted a broadening of the conceptual framework. These theoretical developments reflect the grassroots foundations of environmental justice movements. As movements against environmental injustices gain momentum, they draw upon a wide repertoire of protest strategies to legitimize their claims and broaden their support structure. These strategies range from struggles against distributive injustice, claims for cultural identity, and calls for greater participation in environmental decision-making.

In this chapter, I narrate a story of environmental protest in rural Kentucky and examine the strategies adopted by the community to bolster and gain support for their claims. The narrative examines the evolution of movement discourses on justice in Land Between the Rivers over the course of five decades. The community was uprooted in the mid-twentieth century to make way for dam construction and wildlife preservation and was renamed Land Between the Lakes. The surviving residents and descendants of the community mounted an environmental justice movement and adopted multiple justice imperatives from a plural repertoire of protest strategies to gain legitimacy. While during the early years of the conflict the protestors focused solely on cultural reparations, their attention shifted to economic injustices as the community protested commercialization plans in the wildlife refuge in the mid-1990s. A third shift in the narrative emerged in the mid-2000s when former residents of Land Between the Rivers pressed for greater institutional participation in policy decisions on cultural heritage and environmental protection. Finally, in recent months the former residents have organized parallel heritage tours of Land Between the Rivers as a political protest against the lack of transparency in decision-making. The fluid nature of the narratives attests to two specific issues. First, it captures the dynamic nature of environmental justice movements as protest groups adapt to changing movement imperatives and sociopolitical contexts. Second, by conceptualizing environmental justice through the lens of a plural repertoire of strategies and imperatives, it allows us to incorporate a broad segment of social movements against cultural loss, human rights violations, and unfair regulations within the rubric of environmental justice theory.

Drawing upon the story of the community, I argue that environmental justice movements show remarkable resiliency as communities align their narratives with their contexts to gain legitimacy and build coalitions for success. I apply the four-dimensional framework for environmental justice to explore how protest movements embrace plural tools to mobilize their claims for justice. In this chapter I draw upon the lessons from Land Between the Rivers to explore how the community framed and mobilized their claims for cultural justice, economic justice, institutional justice, and political justice and the specific ways in which sociopolitical and temporal contexts with a movement determined the adoption of one or more of these strategies.

The four-dimensional framework proposes a justice framework that is fluid, multilayered and plural in nature. It allows us to examine how environmental justice discourses evolve over time and adopts multiple narratives as it struggles to gain legitimacy and support. More specifically, the plural protest strategies adopted in the course of the environmental movement in Land Between the Rivers suggests that movement groups are resilient, flexible, and that claims for justice are closely influenced by spatial and temporal contexts within which movements are organized.

The Quest for Justice: Exploring the Protest Strategies in Land Between the Rivers

In this section, I use examples from the case study of an environmental justice movement to illustrate different manifestations of justice strategies that materialize over the lifetime of a movement. This is not intended as a full-blown case study. Instead, I use specific examples to illustrate the contexts within which justice strategies emerge and are adopted as tools of resistance by movement groups. The justice strategies proposed are best studied through the lens of a perception-response duality. In other words, in each case I describe how certain forms of injustices were perceived by the community followed by an analysis of the specific justice strategies used to respond to these injustices. I hope to demonstrate here that justice-seeking in the course of environmental justice movements is plural, strategic, and context-driven. While multiple justice-seeking strategies co-existed during the movement, some gained more prominence due to immediate sociopolitical changes in the region. Additionally some of the justice-seeking strategies bore more dividends than others particularly due to the resonance of the message with the general population, successful coalition building, and public disapproval of the federal proposals.

Figure 5.1 Community protestors, 1964. Image courtesy of David Nickell.

Cultural Justice in Land Between the Rivers

Cultural justice claims emerge in the course of contestations between "expert" and lay knowledge claims. Cultural injustices are perceived by communities when expert knowledge is granted precedence and legitimacy over local "lay" knowledge. For example, TVA's project in Land Between the Rivers was portrayed as a nation-building initiative, designed by scientists whose "expert" knowledge about economic potential of the peninsula was viewed as superior to the residents' cultural claim to their place. This economic vision was promoted and advertised through unjust cultural stereotypes of the local community, selective acknowledgment of local history and heritage, and efforts to transfer management rights over cultural property to "legitimate" agencies. In response, the community collectively developed several strategies to protest the violations of their cultural rights.

TVA's early attempts to rationalize its visions of nation-building were fostered through a dismal portrayal of the living conditions of the community residents in non-modern terms. To garner support for its "development" plans, TVA started a public campaign of stereotypical images of the community as products of a widespread culture of poverty and dysfunctional lifestyles. Community protests were labeled as unnecessary attempts to stem regional economic developmental plans and residents were encouraged to join the nation-building goals of the TVA in order to protect themselves from the staggering poverty of the region. These aggressive advertising initiatives effectively isolated the community from seeking external support for their struggle for cultural justice.

Residents narrate several instances of negative community portrayal by the TVA and shared their collective frustrations with these experiences. Local newspapers shared stories of appalling poverty and abject misery in the communities of Land Between the Rivers. Others highlighted the overuse of forests and land in the peninsula and continued commercialization in the region with little evidence. In fact, documentary evidence provides evidence to the contrary and community's resistance to these portrayals attempted to publicly share these alternate perspectives.

In addition to negative stereotypes, the cultural onslaught of Land Between the Rivers was conducted through selective use of community history. One key example was the lack of attention to the environmental stewardship history of the community as exemplified by the Coalins. Another is the lack of attention to the management and preservation of the remaining church in the land. The community's cultural justice-seeking initiatives are aimed to challenge this non-recognition of their indigenous knowledge claims in current environmental planning.

Finally, the contestation over the management rights of community cemeteries in the refuge led to growing concerns about cultural injustice among the residents. While most of the homesteads and churches that survived the flooding were mostly bulldozed by the TVA, almost two hundred community cemeteries still remain within the federal refuge area and the community retained the burial rights on the grounds. A contestation however emerged in 2001 over the supervision and management of the cemeteries with the transfer of the refuge from the TVA to the Forest Service. One of the first tasks of the new federal agency in 2001 was to streamline refuge management and bring it to full compliance with federal regulations. To this end, the Forest Service proposed a "cemetery handbook" that provided legal guidelines to access and maintain the roughly two hundred family cemeteries that remain in the refuge. As part of the proposed guidelines, the Forest Service decided to create an inventory of cemeteries that was to be documented and evaluated for historical and cultural significance. The community viewed this as a violation of their right to the management of their cemeteries.

For the residents, the cemeteries were the only remaining connection the community shared with their past and an avenue for sustaining their cultural heritage. The proposal to regulate this link was viewed by many residents as an attempt to sever this cultural link. Further, the cemetery handbook, in their view, was in direct violation of the community's agreement with the federal government which stated that the government's rights to the land are superseded by the community's burial rights within their cemeteries (Nickell 2007). Overall, the lack of an open dialogue on this issue was viewed by the former residents as assault on the remaining rights of the community and deemed as an attempt to curtail federally granted privileges.

To contest the injustice, the community argued that the original displacement plans clearly articulate community's right to access and manage their cemeteries. Using the condemnation papers distributed during the displacement era, the community successfully mounted a cultural justice campaign by arguing that their right to use was in compliance with federal cultural heritage acts. Consequently, the Forest Service was forced to admit that while the cemetery ownership remained with the federal government, the community has the absolute right to use (and manage the land). The cemetery handbook was withdrawn by the Forest Service and the community regained its rights to the management of the burial grounds.

One of the important outcomes of the cultural contestation was the growing consensus among the residents for careful documentation and archiving of community's history of settlement and subsequent displacement. The community began this process by creating a local newsletter and other web-based archives in order to document oral histories of early cultural and environmental narratives, disparaged through negative stereotyping and selective omission. The archived documents provide useful insights into the community's attachment to place through the listing of local cultural markers, genealogical details about their history, old family letters, wills, and other documents, narratives documenting history of local environmental activities.

In addition, the residents are also lobbying for a "unique culture" status with the Kentucky State Historic Preservation Office (SHPO) to gain special provisions for management and protection of their cultural heritage. As other justice concerns emerged over the years, the community's efforts to protect their cultural rights allowed them to resist commercialization of their heritage. This was particularly evident during the mid- to late-1990s when the community protested TVA's proposed plans for initiating widespread commercial development of the federal refuge.

Material Justice in Land Between the Rivers: Material justice is defined here as a protest strategy to resist inequality in the distribution of environmental goods and bads. In the case of Land Between the Rivers, material justice concerns emerged when TVA unveiled a plan to turn a shared environmental refuge into a profitable "commercial good." In 1995 TVA announced *The Five Concepts*, a proposal for boosting tourism revenue in the wildlife refuge by promoting commercialization of the land. As part of the plan, TVA proposed a large-scale commercialization project that included development of the 300-mile shoreline along the two lakes created from the Kentucky Dam and the Barkley Dam, construction of marinas, and Disney-style theme parks. The plan was proposed to tap into the socioeconomic development potential of the region and eliminate federal support (Nickell 2007). The proposed plan was a direct violation of the non-contractual agreement between TVA and the former residents in 1965 whereby TVA promised to keep the refuge free from commercialization.

In response to the Five Concepts, the community claims material justice for their community and mounted a protest movement that defined the commercialization attempts as material violations. The response of the community was,

> swift, and TVA held a hearing in response to expressed public concerns. Former residents crowded the meeting hall to decry the development plans as a "total betrayal" of the commitment made by TVA when they "took from us our land and our neighbors"—that is, the Promise. Person after person referred back to the Promise and the "ancestors … buried over there." (Nickell & King n.d.)

The campaign for material justice provided a strong platform for the former residents as they forged durable coalitions with regional and national environmental organizations (like the Audubon Society), hunting groups, and anti-logging activist. Together they formed an organization called the

"Concept Zero Task Force" to vehemently oppose the commercialization plans. *Concept Zero* was a grassroots coalition group composed of the former residents of Land Between the Rivers, local hunting organizations, environmental enthusiasts, and local businesses. The group pressed for continued non-commercialization of the land and garnered enough press coverage to build effective coalitions for environmental justice in the region. As part of their protests the group was able to gather 30,000 signatures protesting the commercialization plans of TVA. The growing public protests led to a Congressional hearing in 1998 and a landmark policy change was initiated as the LBL Protection Act was passed making the "The Promise" legally binding. As part of the directive commercialization plans were halted, the refuge was reverted back to its original state, and the management of the refuge was handed over to the U.S. Forest Service.

An important outcome of the struggle for material justice was that it thrust the movement into the national limelight and captured the public imagination. The community was able to connect their cultural claims with their economic grievances gaining legitimation for their claims. Environmental groups like the Audubon Society, regional conservation societies, Appalachian environmental justice groups, and local hunting and logging interest joined hands with the former residents to protest the commercialization of the refuge. The success of the movement precipitated the community's demand for more robust efforts to encourage greater participation and transparency in environmental decision-making in the refuge—or what I call processual justice.

Processual Justice in Land Between the Rivers

Processual justice is defined here as strategies that promote fairness in institutional and social policies in order to allow the poor and the disenfranchised equal access to social programs and a place at the table in political decision-making. In the case of Land Between the Rivers, demands for processual justice emerged as the former residents sought to consolidate their gains from the anti-commercialization protests. As part of this justice-seeking strategy, former residents called for the creation of a more transparent decision-making structure and were able to gain significant consulting rights in refuge management decision in 2010.

Processual fairness has always been an important roadblock in influencing environmental and cultural policies in Land Between the Lakes. During the displacement residents were often railroaded into accepting lower home prices and denied any institutional recourse. Community residents reported less than half of the assessed value on their land as payment and were discouraged from negotiating higher prices. Residents were often kept in the dark about relocation dates until it was too late to appeal (Nickell 2007). Further, lack of a dialogue in decision-making about the cemetery handbook or the five point plan is also indicative of the lack of institutional transparency and democratic decision-making.

To counter these institutional challenges, former residents adopted two specific strategies. On the one hand, they are working closely with state cultural preservation agencies to gain legal right to influence management of cultural resources in the refuge. Further, residents are working closely with Native American groups, and have contacted the State Historic Preservation Office (SHPO) to attain "Traditional Cultural Place" (TCP) status. TCP status will allow the residents' legal support to work in tandem with the Forest Service in identifying and protecting the cemeteries, church and other cultural markers in the region. As part of this project, community members have been advocating for a place in the National Register. This inclusion will have important institutional justice repercussions. A recent report published by a local community activist in collaboration with a cultural property scholar states this clearly:

> Such recognition does not alter its ownership or management, nor does it convey any property rights to the Between the Rivers people. What it does is offer a strong argument for the community's place at the table when decisions are made by the Forest Service about how the area will be managed. Can the Forest Service still undertake timber sales or recreational development in Between the Rivers? Yes, but as it carries out review of any such action under Section 106 of NHPA, it should consult attentively with the descendant community, and give due consideration to alternatives that will protect the characteristics of the area that the community values. Of course, such consultation can lead to a higher, more integral, more cooperative level of participation by the commu nity. (Nickell & King, draft report n.d: 28)

Second, the community residents applied for and won the right to a consulting party status in the Forest Service Heritage Management Project initiated in 2011 that proposes to protect local cultural heritages and artifacts through predictive modeling and site identification tools. It also proposes to promote values of cultural preservation through environmental and cultural education, monitor site conditions, establish site protocols for discovery of historical sites and remains, and establish theft prevention systems. As an invested actor and a consulting party, the former residents of Land Between the Rivers has a legal claim to participate in consultations about heritage management projects in Land Between the Lakes. While prospects and challenges for participatory justice remain rife, the strategy adopted by the former residents affords their claims an institutional legitimacy.

One outcome of the processual justice strategy is that the former residents were able to offer inputs for cultural heritage plan and share their unique understanding of the history of the land and historical markers that need protection. Second, the former residents have been lobbying with the Forest Service for a more meaningful participatory right through signatory status claims. This claim, offered now to a select few groups, will allow the former residents the ability to approve heritage management decisions by either signing specific initiatives.

In addition to ongoing pursuit of processual fairness, the former residents are also searching for reenergizing coalitions that were successful in blocking commercialization attempts in the late 1990s. Many of these coalitions with extra-local environmental agencies lost traction as commercialization attempts were abandoned in the refuge. To rebuild these networks of justice, the former residents are proposing a set of initiatives to politicize the process and reintroduce it in the public imagination with mixed results. I call these initiatives *political justice* and discuss them below.

Political Justice in Land Between the Rivers

Political justice strategies allow movement groups to seek support through a range of political strategies including practical politicking and appropriation of established movement narratives to legitimize movement claims. In the case of Land Between the Rivers, political justice-seeking has produced some promising coalitions with important implications. For example, the former residents have engaged in vigorous politicking to expand their constituency to gain leverage in negotiations with both the TVA and more recently, with the U.S. Forest Service. Early resistance strategies relied heavily on colorful banner, street-side civil disobedience protests, deputations to political agencies, and signature campaigns. Additionally residents organized local fetes, book fairs, cemetery cleaning drives as a strategy rebuilding community networks and inviting new members and support groups. In recent years, the community has built durable coalitions with local interest groups and environmentalists particularly Heartwood, the Regional Association for Concerned Environmentalists (RACE), and some of the Native American and Eastern Appalachian environmental groups. Additionally, to raise public consciousness about the cultural heritage of Land Between the Rivers, former residents have also organized "unofficial" resident-guided cultural tours of the refuge. These tours allow the residents to share their history with outsiders as well as resist cooptation of their community-initiated projects by the Forest Service. As a community activist pointed out,

> Every time we start a project the Forest Service shuts it down, then in response to the negative publicity they take it over themselves. So we are starting a lot of new projects! They finally gave us consulting party status, after more than ten years of demanding what should have been automatically given to us. But

> they are still trying to keep us at arm's length on anything dealing with "their" heritage program. They claimed the heritage trail at Star Lime Works, and St. Stephens and other projects we had done as major accomplishments of "their" heritage program without ever mentioning us at all. (email communication: community activist, male, age 45)

To resist these appropriations by the U.S. Forest Service, the community has also started annual reunions of past residents at a local school site:

> We are now organizing tours of all the communities, led by people who lived in those communities. We are planning a reunion of Yale School alumni at the school site, which we hope will lead to reunions at the other schools. If they want to take over our projects and claim them as their own, we can just keep coming up with more projects. (community activist, male, age 45)

Additionally, the community has attempted to situate their struggles for continued non-commercialization of the refuge within the broader environmental movement in the United States for greater political success. While similar frame alignment efforts in environmental justice movements were successful, the results so far have been mixed in the case of Land Between the Rivers. A resident lamented about this core challenge:

> We did try to get support, you know, from environmental justice groups in Illinois, groups in Appalachia, Native American groups here in West Kentucky … . But, well, it hasn't really worked wonders. What can we show for ourselves? Our conditions are vastly different from the miners there [in Appalachia]. And we do not have the same legal protections as the Native Americans do.

Another interviewee chalked up the failure in linking their struggle to a relevant theme to the absence of an appropriate label:

> Bottom line is that we have to show some evidence that there really is a Between the Rivers culture that is in any way distinct from the outside culture. The main concern seems to be that there are no clear boundaries [between our status and others] since there is no formally designated culture—us being white and mainstream … as in a Native American tribal registry. So how do we convince anyone that there is coherent group of people that place a different value on the place than what the general population does?

Political justice strategies are recent efforts by the community to seek political support against the injustices suffered by their community. In the immediate future, the community remains resilient in seeking to build enduring political coalitions. First, politicking efforts have produced generally enthusiastic response from local communities and have allowed the former resident sufficient political capital for successfully negotiating both consulting party *and* signatory party status in decisions pertaining to refuge management. Second, the community operated and locally managed tours of the refuge that aim to highlight and share the long history of the community have been effective in granting a measure of control over their own cultural history and heritage. While the tours have no official sanction, they offer the former residents to share their history of displacement and cultural loss with the general public. Third, independent community-initiated projects for cultural heritage preservation have also offered a sense of political agency to the residents and allowed them to seek strategic coalitions with interested parties working toward preservation of the refuge for future generations.

Overall, the four justice strategies demonstrate that justice movements use strategic and context-specific protest strategies to mobilize their claims. Hence, a plural and flexible model of environmental justice presents a more effective perspective to understand and analyze justice movements in action.

Part VI

TOWARD AN INTEGRATIVE FRAMEWORK FOR ENVIRONMENTAL JUSTICE

Concluding Remarks

A small town in the rural South in the United States, Land Between the Rivers remains an idyllic location for visitors intent on reconnecting with nature. It does not immediately strike one as a focal point for debates on links between environmental injustices and cultural identity. Yet, former residents of Land Between the Rivers have been protesting cultural and environmental injustices in their land as they struggle to preserve their identities and their entitlements challenged by displacement and loss due to dam-building and development of an environmental refuge. Their protests highlight the wide range of issues that may emerge in the course of an environmental justice movement as communities aim to resist encroachment on their rights to their land, environmental resources, and cultural identity by the state.

From a theoretical perspective, the case of Land Between the Rivers shows that few environmental justice struggles in communities around the world exhibit only one specific protest strategy (be it distributive, cultural, or institutional); rather many if not most involve plural tools of resistance that are adapted to meet the situation on the ground. The case of land Between the Rivers suggests that a comprehensive and inclusive theory of Environmental Justice should strive to embrace and integrate these multiple ways of representing justice. Toward that end, this book proposes a new framework organized around four justice themes—*cultural, material, processual*, and *political justice*. The first three are derived from scholarly studies in Science and Technology Studies, environmental justice studies, and political philosophy. The fourth theme, *political justice*, refers to the role of politicking in environmental protests to demand social change. It should be clear that the four protest themes are not mutually exclusive and that one theme may continue to operate as a mobilization tool concurrently with other justice strategies.

Over three decades, environmental justice scholars have debated the impact of environmental hazards on poor and marginalized communities. The conceptual richness of this literature can be organized into three complementary scholarly traditions in environmental justice studies, science and technology studies, and political philosophy. For scholars of environmental justice studies, environmental movements refer to grassroots mobilization among communities affected by unfair environmental practices that adversely affect poor communities and communities of color. In other words, EJ movements tend to focus on the *notion of fairness* in distribution of environmental resources. These notions of fairness include unfair perceptions and experiences of environmental distribution, historical and institutional dynamics of injustice, and construction of justice frames in movement narratives (Agyeman 2003; Bryant & Mohai 1992; Pellow 2004; Taylor 2000).

Political philosophers have extensively debated specific political strategies to mitigate injustices by granting *promoting processual or institutional justice* to disenfranchised communities (Rawls 1981, Sen 2000, Schlosberg 2007). These scholars argue that justice is closely linked to the availability of fair institutions and practices that promote fair opportunities to seek redressal and address complaints. A third justice theme emerges from science and technology studies where scholars define the pursuit of environmental justice through adoption of cultural tools of protests that emphasize the role of local communities in shaping environmental and scientific decision-making (Allen 2007; Hess 2007; Frickel & Vincent 2007; Banerjee 2015). The three approaches to environmental justice have flourished in their theoretical richness and methodological diversity, and extensively influenced scholarly literature on the notion of the "justice" in environmental justice.

It is now important to distill the key ideas from these scholarly traditions in order to propose a synthetic framework for environmental justice. This book examines how environmental justice is best conceived as a set of plural protest strategies that are adopted for mobilizing resources in protests. The story of Land Between the Rivers is the case in point; it shows how movement narratives were reframed as needed as key events emerged in the course of the decades.

The environmental justice movement began in the mid-twentieth century when TVA started the construction of the Kentucky and the Barkley dams and removed local residents under federal law, at times forcibly. While originally the removal was justified on grounds of woodlands conservation and protection of water quality, the lakes were never put to use for water supply. Indeed, for years little use was made of the project. However, in recent years the U.S. Forest Service (which has assumed management responsibility since 1999) has begun planning a major recreational project for the region. Outraged by such decisions and sensing a political opportunity in the government agency's implicit recognition that the original dam was unnecessary, the community of Land Between the Rivers has organized a vigorous campaign to mobilize support for continued non-commercial use of the land, and for a return of at least some modest access to the region for the original inhabitants.

Given the long association of the community with their environment, the federal decision to remove the residents from their land created a sense of uprooting, in both a physical and emotional sense. The displacement began with the construction of the Kentucky and the Barkley dams on the rivers surrounding Land Between the Rivers by the Tennessee Valley Authority, who removed local residents under federal law, at times forcibly. While originally the removal was justified on grounds of woodlands conservation and protection of water quality, the lakes were never put to use for water supply. In the early years, the community was impacted in several ways as a result of the displacement. On the one hand, they lost all authority on the forested land that had once been their home, experiencing a shared sense of loss of place and identity. On the other hand, the construction of the refuge led to extensive regulation of the now public land (formerly community land) leading to loss of local cultural resources.

Early protests focused on these discourses on cultural loss as the residents rebuild their lives as they prepared to move away from a land that has been their home for close to two centuries. The environmental justice protests—first against the dam construction and later against the displacement—were largely focused on resisting cultural injustices and concerns over loss of place. Hence, early protest strategies focused on seeking rights to cultural resources, non-commercialization of the land, and continued protection of existing cultural resources including churches, cemeteries, remains of iron furnaces, and so on.

The 1990s however led to a strategic shift in community mobilization. Crippled by extensive reduction in federal money, TVA proposed a new plan in 1995 to promote commercialization in the federal refuge, disregarding its own promise of non-commercialization made during the displacements. Outraged by such decisions and sensing a political opportunity in the government agency's implicit recognition that the original dam was unnecessary, the former residents of Land Between the Rivers mounted a vigorous campaign to mobilize support for continued non-commercial use of the land, and for a return of at least some modest access to the region for the original inhabitants. The protesters adopted an economic/distributive justice theme to resist the commercial exploitation of the refuge and were effective in building key coalitions with hunters, environmentalists, and local communities. The protests against non-commercialization were successful and a congressional hearing led to the transfer of the refuge from the TVA to the United States Forest Service in 1999.

While major commercialization plans were halted by the Forest Service, official attempts to sideline community concerns for their heritage, culture, and identity created discord in the community's relationship with the state agencies. Under the new management, the commercialization plans were largely shelved. New concerns emerged soon over disagreements about several issues including community access rights to cemeteries on the refuge, decision-making about cultural and environmental resources, and increased logging and other forms of commercialization in the land. These new developments led to a third form of protest aimed to gain institution access and representations in decision-making bodies for greater *processual justice* for former residents. Years of continued protests and lobbying for institutional fairness led to official recognition of the community as both a consulting party to environmental and cultural decisions in the LBL refuge management as well as a signatory party to ongoing and future decisions.

A fourth justice theme has also emerged over the course of the movement in Land Between the Rivers. This involves direct politicking through several means to resist displacement and loss of a collective identity for the community. I call this *political justice*. This refers to a "spectrum of tools movement groups use, including practical and ideational political strategies, such as active lobbying, campaigning, politicking, and appropriating established narratives to gain leverage and build constituencies" (Banerjee 2014: 6). It includes a broader and loosely connected array of political maneuvers like strategic alignment with like-minded groups, key connections with experts and scientists, message sharing with local communities, and building a public platform for sharing concerns with a broader audience.

In Land Between the Rivers, the former residents have engaged in several forms of politicking. Early protests directly challenged the TVA narrative of development as a justifiable reason to uproot a community from their homes. Both collective political justice strategies, like street protests, signature campaigns, as well as covert and overt individual resistance were used to resist the violence of displacement. In recent years, the former residents have formulated strategies to garner attention to their cause by organizing community tours of their land that offer opportunities to share local connections to the place with the general audience, a powerful tool to reclaim disappearing local cultural history.

As environmental justice movement spreads globally and gives voice to diverse discourses of justice, it has experienced an internal seismic shift. Environmental protests in the last decade encapsulate this transformation as activists discuss the importance of democracy, human rights, cultural identity, and political participation as integral elements of environmental justice. Theoretical and conceptual analyses are increasingly exploring the multidimensional and shifting course of environmental justice movements as we address questions of pollution, water scarcity, and climate change. Communities have also broadened the scope of environmental justice movement to embrace cultural violations, human rights abuses, climate refugee crises, as well as food disasters. As scholarly research explores these complex narratives of injustice, we need to propose synthetic and inclusive approaches to the study of environmental justice.

In this book, I propose an environmental justice framework that includes four justice themes—*cultural justice*, *economic justice*, *institutional justice*, and *political justice*—as central themes or elements of environmental justice framework. Drawing upon a story of an environmental justice movement in Land Between the Rivers, I explore how environmental activists adopted these justice themes to mobilize community resources and seek support for their initiatives. I argue that while the movement began as a result of cultural and environmental displacement, it quickly adopted an economic injustice framework to protest commercialization plans. Successful protests and a transfer of environmental refuge management led to a third shift in strategy as the former residents lobbied for institutional legitimacy. Finally, recent disappointment with limited inclusion in decision-making has led to adoption of covert and overt politicking strategies to demand political legitimacy. The fluid nature of the discourses suggests that the justice in environmental justice includes more than claims of distributive equity. Instead, narratives of justice are often plural, multilayered, and synthetic strategies that offer legitimacy to environmental protests.

A synthetic, plural justice framework offers several important avenues for future research on environmental justice issues. First, it allows us to conceptualize environmental justice as a synthetic framework that views justice as more than a question of distribution. In fact, environmental justice movements draw upon a diverse repertoire of strategies to gain support and legitimacy for their movement. As communities dealing with disaster rebuild their lives and homes, their diverse needs cannot be captured within an economic/distributive lens. As environmental justice protests transcend national boundaries and are adopted to frame a broad array of movements, conceptual approaches to EJ will need to be more flexible and synthetic. An inclusive integrative framework will allow us to incorporate these diverse narratives and movement strategies to studies of environmental justice.

Second, studies of global environmental justice movements are increasingly focusing on studies of life-cycles of toxic products from its productions to its disposal (Pellow 2007). As movements emerge, coalesce, formalize, and succeed or fail, movement groups adopt and legitimize the use of diverse strategies of protest (Tilly 1978). Land Between the Rivers is a case in point. While early protest insisted on cultural reparations, the movement was formalized and coalitions emerged as the community activists adopted and prioritized their claims against *economic injustices* during the commercialization era. Important strides during this phase also facilitated in lobbying for institutional justice and participation. Finally, limited transparency in decision-making and the community's decision to keep up the pressure has led to adoption of parallel political initiatives for justice. A life-cycle approach to such stories of environmental justice allows us to integrate these diversities into the environmental justice framework.

Third, environmental rights, like right to a pollution-free environment, right to a healthy life, and so on, are most achievable only when the relevant capabilities to function are present (Sen 1992). As we explore the possibilities for an EJ framework, concerns about human rights nonhuman rights all offer interesting and important possibilities for a rights approach to environmental justice. For example, concerns about commons land, non-human and/or animal rights, indigenous rights, cultural property rights including material (artifacts) and non-material (stories, narratives, oral histories, music, and so on) resources offer important possibilities for future research. A synthetic framework allows us to account for the characteristic of the institutional framework for the successful implementation of these rights. We can only call a society just when it has created effective conditions for achieving these capabilities. Hence drawing linkages between rights and environmental justice through capability building will provide a more effective framework for conceptualizing agency within environmental justice theory.

Finally, the struggle for environmental justice in Land Between the Rivers also reflects a broader engagement against cultural and environmental injustice in the United States and elsewhere. These include protests against pipeline, anti-fracking movements, and past and ongoing protests against land grab for developmental projects in developing countries. The ongoing protests organized and led by the Standing Rock Sioux Tribe in North and South Dakota is an important example. The Dakota Access Pipeline Movement (DAPL) is focused on environmental issues like access to fresh drinking water for Native American communities as well as cultural preservation of the tribe's sacred, ancestral land (Shaw 2017; Rodgers & Burleson, 2017; Whyte 2017). These concerns are also reflected in anti-fracking protests in communities across the United States (Brasier 2011; Malin 2014). Extensive protests have also occurred against land grab from disenfranchised communities in developing nations including tribal protests against dam constructions in Latin America, India, and Laos among others (McCormick 2006; Baviskar 2005; Hirsch 2016).

Environmental justice studies have come a long way in the last two decades. Conceived originally as a radical alternative to traditional environmental activism it must now grapple with a new set of environmental challenges emerging from global food crisis, economic downturn, disappearing worker rights, racial politics, cultural inequalities, and climate change. EJ movements have proven its resilience by providing an inclusive platform for all communities affected by global political and economic shifts. Synthetic approaches to examine and study these movements will be key to effectively address specific strategies and possibilities for solving environmental challenges in the twenty-first century.

References

Adamson, J., M. M. Evans, & R. Stein. 2002. *The Environmental Justice Reader: Politics, Poetics, and Pedagogy*. Tucson, AZ: University of Arizona Press.

Agyeman, J., R. D. Bullard, & B. Evans. 2002. "Exploring the Nexus: Bringing Together Sustainability, Environmental Justice and Equity." *Space & Polity*. 6(1): 77–90.

Anderton, D. L., A. B. Anderson, J. M. Oakes, & M. R. Fraser. 1994. "Environmental Equity: The Demographics of Dumping." *Demography*. 31(2): 229–248.

Allen, B. J. 2003. *Uneasy Alchemy: Citizens and Experts in Louisiana's Chemical Corridor Disputes*. Cambridge, MA: MIT Press.

Allen, B. J. 2007. "Environmental Justice and Expert Knowledge in the Wake of a Disaster," *Social Studies of Science*. 37(1): 103–110.

Altman, I. & S. M. Low, (eds.) 1992. *Place Attachment*. New York, NY: Plenum Press.

Bagby, K. & J. Kusel. 2002. *Towards a Civic Science with Undeserved Communities in Community Forestry*. Taylorsville: PWCFC.

Banerjee, D. 2007. "(Re) imagining the Local in Global Environmental Narratives: The Story of Between the Rivers." In S. Dasgupta (ed.) *Understanding Global Environments*. New Delhi, India: Sage Publications.

Banerjee, D. 2014. "Toward an Integrative Framework for Environmental Justice Research: A Synthesis and Extension of the Literature." *Society and Natural Resources*. 27(8): 805–819.

Banerjee, D. & S. L. Steinberg. 2015. "Exploring Spatial and Cultural Discourses in Environmental Justice Movements: A Study of Two Communities." *Journal of Rural Studies*. 39 (June): 41–50.

Bath, R. C., J. Tanski, & R. Villareal. 1994. "The Politics of Water Allocation in El Paso County Colonies." *Journal of Borderlands Studies*. 9(1): 14–38.

Baviskar, A. 2005. *In the Belly of the River: Tribal Conflicts over Development in the Narmada Valley*. Oxford, U.K.: Oxford University Press.

Beck, U. 1992. *Risk Society: Towards a New Modernity.* (trans. by Mark Ritter). Newbury Park, CA: Sage Publications.

Beck, U. 1992. *Risk Society -Towards a New Modernity.* London, UK: Sage Publications.

Been, V. 1994. "Locally Undesirable Land Uses in Minority Neighborhoods: Disproportionate Siting or market Dynamics?" *Yale Law Journal.* 103(6): 1383–1422.

Bell, M. M. 1994. *Childerley: Nature and Morality in a Country Village.* Chicago, IL: University of Chicago Press.

Bell, M. M. & M. S. Carolan. 2004. *An Invitation to Environmental Sociology.* 2nd edition. Thousand Oaks, CA: Pine Forge Press (Sage).

Berry, B. L. (ed.) 1977. *The Social Burden of Environmental Pollution: A Comparative Metropolitan Data Source.* Cambridge, MA: Ballinger Publishing.

Blum, E. 2008. *Love Canal Revisited: Race, Class, and Gender in Environmental Activism.* Lawrence: University of Kansas Press.

Bourdieu, P. 1992. *The Logic of Practice.* Stanford, CA: Stanford University Press.

Bowen, W. M., M. J. Salling, K. E. Haynes & E. J. Cyran. 1995. "Towards Environmental Justice: Spatial Equity in Ohio and Cleveland." *Annals of the Association of American Geographers.* 85(4): 641–663.

Brasier, K., M. Filteau & D. McLaughlin. 2011. "Residents' perceptions of community and environmental impacts from development of natural gas in the Marcellus Shale: a comparison of Pennsylvania and New York cases." *Journal of Rural Social Sciences.* 26(1): 32–61.

Bridger, J. 1996. "Community Imagery and the Built Environment." *The Sociological Quarterly.* 37(3): 353–374.

Bryant, B. 1995. *Environmental Justice: Issues, Policies and Solutions.* Washington, DC: Island Press.

Bryant, B. & P. Mohai. 1992. *Race and the Incidence of Environmental Hazards.* Boulder, CO: Westview Press.

Bryant, B. & P. Mohai. (eds.) 1992. *Race and the Incidence of Environmental Hazards: A Time for Discourse.* Boulder, CO: Westview Press.

Bullard, R. D. 1983. "Solid Waste Sites and the Houston Black Community." *Sociological Inquiry.* 53(Spring): 273–288.

Bullard, R. D. 1983. "Solid Waste Sites and the Black Houston Community." *Sociological Inquiry*, 53(2–3): 273–288.

Bullard, R. D. (ed.) 1993. *Confronting Environmental Racism: Voices from the Grassroots.* Boston: South End.

Bullard, R. D. 1994a. *Dumping in Dixie: Race, Class, and Environmental Quality.* 2nd ed. Boulder, CO: Westview.

Bunker, S. 1985. *Underdeveloping the Amazon: Extraction, Unequal Exchange, and the Failure of the Modern State.* Urbana, IL: University of Illinois Press.

Burch, W. R. Jr. 1976. "The Peregrine Falcon and the Urban Poor: Some Sociological Interrelations." In P. Richardson & J. McEvoy (eds.) *Human Ecology: An Environmental Approach.* Pp. 308–316. Belmont, CA: Duxbury Press.

Capek, S. M. 1993. "The 'Environmental Justice' Frame: A Conceptual Discussion and an Application." *Social Problems.* 40: 5–24.

Carson, R. 1962. *Silent Spring*. Boston, MA: Houghton Mifflin.
Carter, E. D., B. Silva, & G. Guzman. 2011. "Migration, Acculturation, and Environmental Values: The Case of Mexican Immigrants in Central Iowa." *Annals of the Association of American Geographers*. 103(1): 129–147.
Catton, W. R. Jr. 1964. "The Development of Sociological thought." In R. E. L. Farris (ed.) *Handbook of Modern Sociology*. Pp. 912–950. Chicago, IL: Rand McNally.
Chambers, R. 1994. "Participatory Rural Appraisal (PRA): Challenges, Potentials and Paradigm." *World Development*. 22(10): 1437–1454.
Churchill, C. J. 2005. "Ethnography as Translation." *Qualitative Sociology*. 28(1): 3–24
Coffey, A. 1999. *Ethnographic Self: Fieldwork and the Representation of Identity*. London, UK: Sage Publications.
Collins, L. 1847. *History of Kentucky*. Cincinnati, OH: Lewis Collins and J.A. Collins.
Comancho, D. E. (ed.) 1998. *Environmental Injustices, Political Struggles: Race, Class, and the Environment*. Durham, NC: Duke University Press.
Daniels, A. K. 1967. "The Low-caste Stranger in Social Research." In G. Sjoberg (ed.) *Ethics, Politics, and Social Research*. Pp. 267–296. Cambridge, MA: Schenkman Publishing.
Darier, E. (ed.) 1999. *Discourses of the Environment*. Oxford: Blackwell.
Davidman, L. 1997. "The Personal, the Sociological, and the Intersection of the Two." *Qualitative Sociology*. 20(4): 507–515.
Demaria, F. 2010. "Shipbreaking at Alang-Sosiya: An Ecological Distribution Conflict." *Ecological Economics*. 70(2): 250–260.
DeVault, M. 1997. "Are We Alone?" *Qualitative Sociology*. 20(4): 499–506.
Dobson, A. (ed.) 1999. *Fairness and Futurity: Essays on Environmental Sustainability and Social Justice*. New York: Oxford University Press.
Du Bois, W. E. B. 1997. *The Souls of Black Folk*. Boston, MA: Bedford Books.
Duneier, M. 1999. *Sidewalk*. Farrar, Strauss and Giroux. New York, NY.
Dupuis, E. M. & P. Vandergeest (eds.) 1996. *Creating the Countryside: The Politics of Rural and Environmental Discourse*. Philadelphia, PA: Temple University Press.
Dworkin, R. M. 2000. *Sovereign Virtue: The Theory and Practice of Equality*. Cambridge, MA: Harvard University Press.
Ehrlich, P. 1968. *The Population Bomb*. New York: Ballantine.
Elder, G. H., V. King & R. D. Konger. 1996. "Attachment to Place and Migration Prospects: A Development Perspective." *Journal of Research on Adolescence*. 6: 397–425.
Emerson, R. W. 1985 [1849]. *Nature: Addresses and Lectures*. New York, NY: Library of America.
Epstein, B. 1997. "The Environmental Toxics Movement: Politics of Race and Gender." *Capitalism, Nature, Socialism*. 8: 63–87.
Feld, S. & H. B. Keith (eds.) 1996. *Senses of Place*. Santa Fe, NM: School of American Research Press.
Ferguson, J. 1994. *The Anti-Politics Machine: 'Development,' Depoliticisation and Bureaucratic Power in Lesotho*. Minneapolis, MN: University of Minnesota Press.
Fischer, F. 2000. *Citizens, Experts, and the Environment: The Politics of Local Knowledge*. Durham, NC: Duke University Press.

Fontana, A. & T. A. McGinnis. 2005. "The Strangeness of Being." *Studies in Symbolic Interaction*. 28: 205–216.

Foresta, R. A. 2013. *Land Between the Lakes: A Geography of the Forgotten Future*. Knoxville, TN: University of Tennessee Press.

Fortmann, L. 1995. "Talking Claims: Discursive Strategies in Contesting Property." *World Development*. 23(6): 1053–1063.

Foucault, M. 1984. *The Foucault Reader*. P. Rainbow (ed.) New York, NY: Pantheon Books.

Freeman, A. M. III. 1972. "The Distribution of Environmental Quality." In A. V. Kneese & B. T. Bower (eds.) *Environmental Quality Analysis: Theory and Method in the Social Sciences*. Pp. 243–278. Baltimore: Johns Hopkins University Press.

Frickel, S. & M. B. Vincent. 2007. "Katrina, Contamination, and the Unintended Organization of Ignorance." *Technology in Society*. 29: 181–188.

Friedman-Jimenez. G. 1989. "Occupational Disease among Minority Workers; A Common and Preventable Public Health Problem." *American Association of Occupational Health Nurses Journal*. 37(2): 64–70.

Friedman, H. & P. McMichael. 1989. "Agriculture and the State System: The Rise and Decline of National Agriculture—1870 to present." *Sociologia Ruralis*. 39: 93–117.

Gans, H. J. 1982. *Urban Villagers: Group and Class in the Life of Italian Americans*. New York: Free Press.

Giddens, A. 1984. *The Constitution of Society: Outline of the Theory of Structuration*. Cambridge, UK: Polity Press.

Gieryn, T. F. 2000. "A Space for Place in Sociology." *Annual Review of Sociology*. 26: 463–496.

Gottlieb, A. & P. Graham. 1996. "Choosing a Host." In A. Lareau & J. Shultz (eds.). *Journeys Through Ethnography: Realistic Accounts of Fieldwork*. Pp. 75–110. Boulder, Co: Westview Press.

Gottlieb, R. 2005. *Forcing the Spring: The Transformation of the American Environmental Movemen*t. Washington: Island Press.

Gould, K. A., A. Schnaiberg, & A. S. Weinberg. 1996. *Local Environmental Struggles*. New York: Cambridge University Press.

Gould, K. A., D. N. Pellow, & A. Schnaiberg. 2004. "Interrogating The Treadmill Of Production: Everything You Wanted To Know About The Treadmill, But Were Afraid To Ask." *Organization & Environment*. 17(3): 296–316.

Grieder, T. & L. Garkovich. 1994. "Landscapes: The Social Construction of Nature and the Environment." *Rural Sociology*. 59: 1–24.

Guha, R. 1989. "Radical American Environmentalism & Wilderness Preservation: A Third World Critique." *Environmental Ethics*. 11: 71–83.

Guha, R. 1990. *The Unquiet Woods: Ecological Change and the Peasant Rebellion in the Himalayas*. Berkeley, CA: University of California Press.

Guha, R. 2000. *Environmentalism: A Global History*. New York: Longman.

Gupta, A. & J. Ferguson. 1992. "Beyond 'Culture': Space, Identity and the Politics of Difference." *Cultural Anthropology*. 7(1): 6–23.

Gupta, A. & J. Ferguson. 1997. "Discipline and Practice: The Field as Site, Method, and Location in Anthropology." In A. Gupta & J. Ferguson (eds.) *Anthropological Locations: Boundaries and Grounds of a Field Science*. Pp. 1–46. Berkeley, CA: University of California Press.

Halbwachs, M. 1992. *On Collective Memory* (ed. & trans. L. Coser). New York: Harper & Row.

Hamer, P. B. 1933. *Tennessee: A History 1673–1932*. New York: The American Historical Society, Inc. Vol 1.

Harvey, D. 1996. *Justice, Nature and the Geography of Difference*. Cambridge, MA: Blackwell.

Harvey, D. 2000. *Spaces of Hope*. Berkeley, CA: University of California Press.

Heiman, M. K. 1996. "Race, Waste, and Class: New Perspectives on Environmental Justice." (Editor's Introduction for a Special Edition of Antipode). *Antipode*. 28(2): 111–121.

Henry, J.M. 1975. *Land Between the Rivers*. Clarksville, TN: Austin Peay University Press.

Hirsch, P. 2016. "The Shifting Regional Geopolitics of Mekong Dams." *Political Geography*. 51: 63–74.

Hogan, W. T. 1971. *An Economic History of the Iron and Steel Industry in the United States*. Washington, DC: Heath and Company [Vol 1].

Hooks, G. & C. L. Smith. 2004. "The Treadmill of Production: National Sacrifice Areas and Native Americans." *American Sociological Review*. 69: 558–575.

Hummon, D. M. 1992. "Community Attachment: Local Sentiment and Sense of Place." Pp. 253–278. In I. Altman & S. M. Low (ed.) *Place Attachment*, New York: Plenum Press.

Irwin, A. 2001. "Constructing the Scientific Citizen: Science and Democracy in the Biosciences." *Public Understanding of Science*. 10(1): 1–18.

Jasanoff, S. 1992. "Science, Politics and the Renegotiation of Expertise." *Osiris*. 7: 1–23.

Katz, J. 1997. "Ethnography Warrants." In *Sociological Methods and Research*. 25(4): 391–412.

Keith, M. & S. Pile (eds.) 1993. *Place and the Politics of Identity*. London, UK: Routledge.

Krieger, S. "Beyond Subjectivity." In A. Lareau & J. Shultz (eds.). *Journeys through Ethnography: Realistic Accounts of Fieldwork*. Pp. 177–194. Boulder, Co: Westview Press.

Labalme, J. 1988. "Dumping on Warren County. Environmental Politics." In B. Hall (ed.) *Environmental Politics: Lessons from the Grassroots*. Pp. 23–30. Durham, NC: Institute for Southern Studies.

Lafferty, M. W. 1971. *The Lure of Kentucky*. Detroit, MI: Singing Tree Press.

Lavelle, M. & M. Coyle. 1992. "Unequal Protection: The Racial Divide in Environmental Law." *National Law Journal*. 15(3): S1–S16.

Lawrence, D. L. & S. M. Low. 1990. "The Built Environment and Spatial Form." *Annual Review of Anthropology*. 19: 453–505.

Lerner, S. 2005. *Diamond: A Struggle for Justice in Louisiana's Chemical Corridor*. Cambridge, MA: The MIT Press.

Low, S. M. 1992. "Symbolic Ties that Bind: Place Attachment in the Plaza." In Irwin & Low (eds.) *Place Attachment*. New York: Plenum Press.

Lukes, S. 1993. "Five Fables about Human Rights." In S. Hurley & S. Shute (eds.) *On Human Rights*. New York, NY: Basic Books.

Lund, C. 2001. "Precarious Democratization and Local Dynamics in Niger: Micro-politics in Zinder." *Development and Change*. 32: 845–869.

Maines, D. 1993. "Narrative's Moment and Sociology's Phenomena: Towards a Narrative Sociology." *The Sociological Quarterly*. 34: 17–38.

Malin, S. 2014. "There's no real choice but to sign: neoliberalization and normalization of hydraulic fracturing on Pennsylvania farmland." *Journal of Environmental Studies and Sciences*. 4(1): 17–27.

Malinowski, B. 1961. *Argonauts of the Western Pacific: An Account of Native Enterprise and Adventure in the Archipelagoes of Melanesian New Guinea*. New York: Dutton.

Marbury, H. J. 1995. "Hazardous Waste Exportation: The Global Manifestation of Environmental Racism." *Vanderbilt Journal of Transnational Law*. 28: 1225–1237.

Martinez Alier, J. 2003. "Problems of Ecological Degradation: Environmental Justice or Ecological Modernization?" *Capitalism, Nature, Socialism*. 14(1): 133–138.

McClain, I. H. 1965. *A History of Stewart County, Tennessee*. Columbia, TN: Iris Hopkins McClain.

McCormick, S. 2006. "The Brazilian Anti-Dam Movement: Knowledge Contestation as Communicative Action." *Organization & Environment*. 19(3): 321–346.

McLemore, S. D. 1970. "Simmel's 'Stranger': A Critique of the Concept." *Pacific Sociological Review*. 13(2): 86–94.

McWhiney, G. 1988. *Cracker Culture and Celtic Ways in the Old South*. Tuscaloosa, AL: University of Alabama Press.

Miller, D. 1999. "Social Justice and Environmental Goods." In A. Dobson (ed.) *Fairness and Futurity: Essays on Environmental Sustainability and Social Justice*. Pp. 151–172. New York: Oxford University Press.

Montell, W. L. 1975. *Ghosts along the Cumberland: Deathlore in the Kentucky Foothills*. Knoxville, TN: University of Tennessee Press.

Muir, J. 1901. *Our National Parks*. Houghton: Mifflin.

Nash, D. J. & A. V. Wolfe. 1957. "The Stranger in Laboratory Culture." *American Sociological Review*. 22(4): 400–405.

Nickell, D. 2001. "Land Between the Lakes: The Story of Colonialism in Kentucky." Network News, *Sociologists for Women in Society*. XVIII(1): 1–10. Retrieved August 3, 2006, from http://nickell.tierranet.com/tales/lbl.pdf

Nickell, D. & T. F. King. 2004. "Traditional Cultural Places and Rural Historic Landscapes: The Case of 'Between the Rivers.'" Unpublished Manuscript.

Nickell, D. L. 2007. "Between the Rivers: A Socio-Historical Account of Hegemony and Heritage." *Humanity & Society*. 31: 164–209.

Nisbet, R. A. 1959. "The Contributions of Georg Simmel: Comment." *American Sociological Review*. 24(4): 479–481.

Norton, B. & B. Hannon. 1997. "Environmental Values: A Place-based Approach." *Environmental Ethics*. 19(4): 227–246.

Nussbaum, M. C. 2000. *Women and Human Development*. Cambridge, UK: Cambridge University Press.

Ottinger, G. 2010. "Buckets of Resistance: Standards and the Effectiveness of Citizen Science." *Science, Technology, and Human Values*. 35(2): 244–270.

Paige, H. 1996. "Leave If You Can." Pp. 11–14. In W. Vitek & W. Jackson (eds.) *Rooted in the Land: Essays on Community and Place*. New Haven, CT: Yale University Press.

Parks, J. H. & S. J. Folmsbee. 1973. *The Story of Tennessee*. Normal, OK: Harlow Publishing Corporation.

Paulsen, K. E. 2004. "Making Character Concrete: Empirical Strategies for Studying Place Distinction." *City and Community*. 3(3): 243–262.

Pellow, D. N. 2000. "Environmental Inequality Formation: Toward a Theory of Environmental Injustice." *American Behavioral Scientist*. 43: 581–601.

Pellow, D. N. 2002. *Garbage Wars: The Struggle for Environmental Justice in Chicago*. Cambridge, MA: The MIT Press.

Pellow, D. N. 2004. "The Politics of Illegal Dumping: An Environmental Justice Framework." *Qualitative Sociology*. 27(4): 511–525.

Pellow, D. N. & H. N. Brehm. 2013. "An Environmental Sociology for the 21st Century." *Annual Review of Sociology*. 39: 229–250.

Peluso, N. L. 1992. *Rich Forests, Poor People: Resource Control and Resistance in Java*. Berkeley, CA: University of California Press.

Pfiefer, G. M. 2016. "Pesticides, Migrant Farm Workers, and Corporate Agriculture: How Social Work Can Promote Environmental Justice." *Journal of Progressive Human Services*. 27(3): 175–190.

Pickering, A. 1992. "From Science as Knowledge to Science as Practice." A. Pickering (ed.), *Science as Practice and Culture*. Chicago, IL: University of Chicago Press.

Pierce, J. 2003. "Special Issue Introduction." *Qualitative Sociology*. 26(3): 307–312.

Poore, D. 2006. "Durbin, Obama seek U.S. Funds for 'Land Between the Rivers.'" St. Louis Post-Dispatch. Retrieved May 24, 2006 from Sen. Barack Obama's Online Official Website: http://obama.senate.gov/news/060524-durbin_obama_seek_us_funds_for_land_between_the_rivers/index.html

Pulido, L. 1996. "A Critical Review of the Methodology of Environmental Racism Research." *Antipode*. 28(2): 142–159.

Rabinow, P. (ed.) 1984. *The Foucault Reader*. New York: Pantheon Books.

Rawls, J. 1971. *A Theory of Justice*. Oxford: Oxford University Press.

Rawls, J. 1995. *Political Liberalism*. New York: Columbia University Press.

Rawls, J. 2001. *Justice as Fairness: A Restatement*. Edited by E. Kelly. Cambridge, MA: Harvard University Press.

Redclift, M. 1987. *Sustainable Development: Exploring the Contradictions*. London: Methuen.

Rodgers, W. H. Jr. & E. Burleston. 2017. "Dakota Access to Justice and Pipeline Politics: Tribal Consultation, Environmental Justice and Rules of Engagement." *Environmental Law Treatise 2nd Ed.* https://ssrn.com/abstract=2954308 (Accessed on June 9, 2017).

Rotenberg, R. & G. McDonogh. 1993. *The Cultural Meaning of Urban Space*. London: Bergin & Garvey.

Schlosberg, D. 2004. "Reconceiving Environmental Justice: Global Movements and Political Theories." *Environmental Politics*. 13(3): 517–540.

Scholsberg, D. 2007. *Defining Environmental Justice*. Oxford, UK: Oxford University Press.

Schnaiberg, A., K. A. Gould, & D. Pellow. 2004. "Interrogating the Treadmill of Production: Everything You Wanted to Know About the Treadmill, But Were Afraid To Ask." *Organization & Environment*. 17(3): 296–316.

Schram, K. 2005. "You Have Your Own History, Keep Your Hands Off Ours: On Being Rejected in the Field." *Social Anthropology*. 13(2): 171–183.

Scott, J. C. 1985. *Weapons of the Weak: Everyday Forms of Peasant Resistance*. New Haven, CT: Yale University Press.

Scott, J. C. 1998. *Seeing Like a State: How Certain Schemes to Improve the Human Condition Have Failed.* New Haven, CT: Yale University Press.

Scott, R. R. 2010. *Removing Mountains: Extracting Nature and Identity in the Appalachian Coalfields*. Minneapolis, MN: University of Minnesota Press.

Selznick, P. 1949. *TVA and the Grassroots: A Study in the Sociology of Formal Organization*. Berkeley, CA: University of California Press.

Selznick, P. 1996. "In Search of Community." In W. Vitek & W. Jackson (eds.) *Rooted in the Land: Essays on Community and Place*. Pp. 195–206. New Haven, CT: Yale University Press.

Sen, A. K. 1981. *Poverty and Famines: An Essay on Entitlement and Deprivation.* New York and Oxford: Oxford University Press.

Sen, A. K. 1992. *Inequality Reexamined.* Cambridge, MA: Harvard University Press.

Sen, A. K. 1999. *Development as Freedom.* New York: Anchor Books.

Sen, A. K. 2002. *Rationality as Freedom*. Cambridge, MA: Harvard University Press.

Sen, A. K. 2006. *Identity and Violence: The Illusion of Destiny*. New York: W. Norton & Co.

Shaw, R. J. 2017. What Standing Rock Teaches Us About Environmental Racism And Justice. http://healthaffairs.org/blog/2017/04/17/what-standing-rock-teaches-us-about-environmental-racism-and-justice/ (Accessed on June 9, 2017).

Sierra Club. 2001. *Sierra Club Conservation Policies*. Retrieved August 3, 2006 from the Sierra Club Website: http://www.sierraclub.org/policy/conservation/justice.asp

Silliman, J. 1997. "Making the Connections. Women's Health and Environmental Justice." *Race, Gender, and Class*. 5: 104–129.

Simmel, G. 1950 [1908]. "The Stranger." In K. Wolff (trans.) *The Sociology of Georg Simmel*. Pp. 402–408. New York: Free Press.

Simms, B. 2007. "Things Fall Apart: Disaster, Infrastructure, and Risk." *Social Studies of Science*. 37(1): 93–95.

Sinclair, U. 1906. *The Jungle*. Mineola, NY: Dover Publications.

Stevens, S. M. 2007. "Speaking Out: Toward an Institutional Agenda for Refashioning STS Scholars as Public Intellectuals." *Science, Technology & Human Values*. 33(6): 730–753.

Stewart, K. 1996. *A Space on the Side of the Road: Cultural Poetics in an Other America*. Princeton, NJ: Princeton University Press.

Stubblefield, F. A. 1968. "Submission before House of Representatives, Subcommittee on Flood Control of the Committee of Public Works." Ninetieth Congress, Second Session, May 15, 1968, Washington, DC.

Szasz, A. 1994. *Ecopopulism: Toxic Waste and the Movement for Environmental Justice*. Minneapolis, MN: University of Minnesota Press.

Szasz, A. & M. Meuser. 1997. "Environmental Inequalities: Literature Review and proposals for New Directions in Research and Theory." *Current Sociology*. 45(3): 99–120.

Szasz, A. & M. Meuser. 1998. "Unintended, Inexorable: The Production of Environmental Inequalities in Santa Clara County, California." *American Behavioral Scientist*. 43: 602–632.

Sze, J. & J. London. 2008. "Environmental Justice at the Crossroads." *Sociology Compass*. 2(4): 1331–1354.

Taylor, D. E. 1993. "Environmentalism and the Politics of Inclusion." In R. Bullard (ed.) *Confronting Environmental Racism: Voices from the Grassroots*. Boston: South End Press.

Taylor, D. E. 2000. "The Rise of the Environmental Justice Paradigm." *American Behavioral Scientist*. 43(4): 508–579.

Taylor, D. E. 2009. *The Environment and the People in American Cities, 1600s–1900s*. Durham, NC: Duke University Press.

Tennessee Valley Authority. 1983. *A History of the Tennessee Valley*. Knoxville, TN: Tennessee Valley Authority Information Office.

Tuan, Y-F. 1993. *Passing Strange and Wonderful: Aesthetics, Nature, and Culture*. Washington, DC: Island Press.

United Church of Christ. 1987. *Toxic Wastes and Race in the United States*. New York: United Church of Christ Commission for Racial Justice.

Urkidi, L. & M. Walter. 2011. "Dimensions of Environmental Justice in anti-gold mining movements in Latin America." *Geoforum*. 42(6): 625–732.

U. S. General Accounting Office. 1983. *Siting of Hazardous Waste Landfills and Their Correlation with Racial and Economic Status of Surrounding Communities*. GAO/RCED-83–168. Washington, DC: Government Printing Office.

USFS Report on the Homeplace. (n.d.). Retrieved October 31, 2017, from https://www.landbetweenthelakes.us/seendo/attractions/homeplace/

USFS Report. 2004a. *Land and Resource Management Plan*. Retrieved August 3, 2006, from the LBL Official Website: http://www.lbl.org/ADMIN/FinalAreaDocumentsMaps/4_PLAN.PDF

USFS Report. 2004b. *Response to Public Concern Report Supplement to Appendix A of the FEIS: Final Environmental Impact Statement Land Between The Lakes National Recreation Area*. Retrieved August 11 2006, from LBL Official website: http://www.lbl.org/ADMIN/FinalAreaDocumentsMaps/8_public_concern_response.pdf

USFS Iron Ore Industry Report (n.d.). Retrieved October 31, 2017, from https://www.landbetweenthelakes.us/seendo/self-guided-activites/iron/

USFS Report. 2006a. *Consolidated Decision for Appeals of the Land Between the Lakes Land and Resource Management Plan.* Retrieved August 11, 2006, from US Forest Service Website: http://www.fs.fed.us/emc/applit/includes/lbl_decision.pdf

USFS *LBL Maps.* 2006b. Retrieved August 19, 2006, from USFS website: http://www.lbl.org/VCMaps.html

Walker, G. 2009. "Contextualization and Evolution Globalizing Environmental Justice: The Geography and Politics of Frame." *Global Social Policy.* 9(3): 355–382.

Wallace, B. J. 1992. *Between the Rivers: History of the Land Between the Lakes.* Clarksville, TN: Austin Peay State University, The Center for Field Biology.

Walter, E. V. 1988. *Placeways: A Theory of the Human Environment.* Chapel Hill: University of North Carolina Press.

Weinberg, A. S. 1997. "Local Organizing for Environmental Conflict: Explaining Differences Between Cases of Participation and Nonparticipation." *Organization & Environment.* 10(2): 194–216.

Weinberg, A. 1998. "The Environmental Justice Debate: A Commentary on Methodological Issues and Practical Concerns." *Sociological Forum.* 13: 25–32.

Wells, M. 1996. *Strawberry Fields: Politics, Class and Work in California Agriculture.* Ithaca, NY: Cornell University Press.

Westra, L. & B. Lawson. 2001. *Faces of Environmental Racism: Confronting Issues of Global Justice.* Lanham, MD: Rowman & Littlefield Publishers.

Whyte, W. F. 1955. *Street Corner Society: The Social Structure of an Italian Slum.* Chicago, IL: University of Chicago Press (2nd edition).

Whyte, W. F. 1999. "On the Evolution of the Street Corner Society. In A. Lareau & J. Shultz (eds.). *Journeys Through Ethnography: Realistic Accounts of Fieldwork.* Pp. 111–173. Boulder, CO: Westview Press.

Whyte, K. 2017. "The Dakota Access Pipeline, Environmental Injustice, and U.S. Colonialism." *Red Ink: An International Journal of Indigenous Literature, Arts, & Humanities.* 19(1): https://ssrn.com/abstract=2925513 (Accessed on June 9, 2017).

Williams, D. R. & D. S. Carr. 1993. "The Sociocultural Meaning of Outdoor Recreation Places." In Ewert et al. (eds.) *Culture, Conflict and Communication in the Wildlife-Urban Interface.* Boulder, CO: Westview Press.

Williams, D. R. & S. L. Stewart. 1998. "Sense of Place: An Elusive Concept that is Finding a Home in Ecosystem Management." *Journal of Forestry.* 96(5): 18–23.

Wood, M. M. 1934. *The Stranger: A Study in Social Relationships.* New York: Columbia University Press.

Yearley, S. 1996. *Sociology, Environmentalism and Globalization: Reinventing the Globe.* London, UK: Sage Publications.

Zimmerman, R. 1993. "Social Equity and Environmental Risk." *Risk Analysis: An International Journal.* 13(6): 649–666.

Zupan, J. M. 1973. *The Distribution of Air Quality in the New York Region.* Baltimore: Johns Hopkins University Press.

Index

About the Author

Damayanti Banerjee is a research faculty affiliate at Colorado State University and received her PhD in Sociology from the University of Wisconsin-Madison. Her research and scholarship examines environmental justice movements, environmental sociology, and energy policy and governance. She has extensively published on place, culture, and environmental identity, environmental mobilization in local communities, and science-making and environmental governance in hydraulic fracturing policies in the United States.

Lightning Source UK Ltd.
Milton Keynes UK
UKHW02n0253191217
314738UK00013B/485/P